AF266852

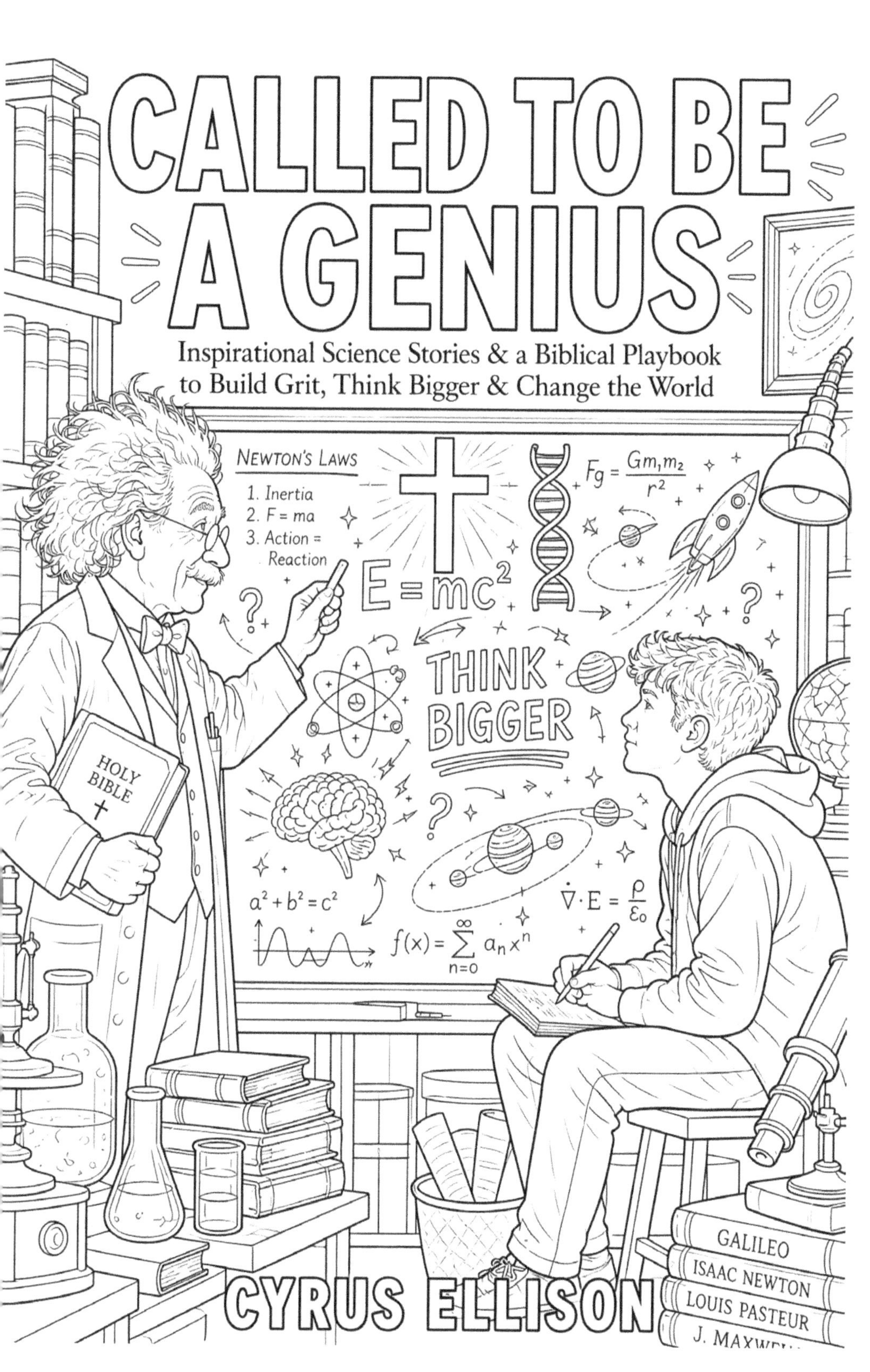

CALLED TO BE A GENIUS
Inspirational Science Stories & a Biblical Playbook
to Build Grit, Think Bigger & Change the World
NEWTON'S LAWS
1. Inertia
2. F = ma
3. Action = Reaction
THINK BIGGER
HOLY BIBLE
GALILEO
ISAAC NEWTON
LOUIS PASTEUR
J. MAXWELL
CYRUS ELLISON

PRAISE

Sandra and Thomas P., Parents of a 13-Year-Old

"Our daughter is brilliant and she knows it. That was starting to become a problem. She was getting arrogant in a way that made us nervous. We gave her this book half hoping it would humble her and half hoping it would channel her. It did both. She came to us after the Florence Nightingale chapter and said she had never thought about how much courage it takes to be right when everyone around you is wrong and still stay humble. She is easier to be around. More curious. Less interested in proving she is the smartest person in the room and more interested in actually solving things. That shift came from this book."

Elijah M., 10th Grader and Aspiring Engineer

"I was failing physics and seriously thinking about giving up on anything science-related. My grades said I was not built for it. This book made me feel insane in the best way. Gregor Mendel failed his exams twice. His supervisors thought he was wasting their time. He died without anyone knowing what he had done. Then 16 years later, the entire world of genetics ran on his work. I read that chapter three times. I kept thinking: what if I quit one year before my peas start making sense?

I have not quit. My grade is up 18 points.
I do not think those things are unrelated."

Ms. Tina V., STEM Teacher and Homeschool Co-op Instructor

"I have been looking for a resource that takes the science seriously and the faith seriously at the same time. Most books do one or the other. This one does both without apology. The chapters are structured well, the stories are compelling, and the Bible connections are not decorative. They are load-bearing. My students who already loved science loved learning about the people behind the discoveries. My students who struggled with science found themselves in the stories of failure, doubt, and persistence. I have ordered copies for every teacher in our co-op."

Pastor Jonathan B., Youth Pastor and Campus Ministry Leader

"I work with teenagers who have been told they have to choose between their faith and their intellect. I give them this book and the choice disappears. It is that simple. Thirty scientists who were Christians, not despite their curiosity but because of it. The chapter on George Washington Carver alone is worth the price. He prayed before every session in his lab and credited God for every discovery. And then he turned down Thomas Edison. That combination of faith and excellence, of humility and total confidence in what God put in you, is exactly what I want young people to see. This book shows it without flinching."

For every kid who has ever stared at something broken, beautiful, or confusing, and couldn't stop wondering why.

This book exists because of a lie.

The lie is this: scientists are a special kind of person. Smarter than the rest. Born different. Gifted from birth in ways that ordinary people aren't. You either have it or you don't.

Every scientist in this book would tell you that is completely wrong.

What they had was not a special brain. It was a question they couldn't let go of. A problem that showed up in their minds first thing in the morning and last thing at night. A stubbornness that refused to accept "that's just how it is."

And almost every one of them had faith. Not faith that made them doubt science. Faith that made them better scientists. Faith that told them the universe was real, ordered, knowable, and worth studying. Faith that kept them going when the experiments failed, when the critics laughed, when the work seemed pointless.

These are their stories. Not the polished, finished, famous versions. The real ones. The versions where they were wrong, afraid, ignored, and broke. The versions where everything depended on one more question.

Your curiosity is not an accident. It was put there on purpose. Use it.

YOUR EXPERIMENT STARTS NOW

Nobody told you this in school.

The scientists who changed the world, the ones whose names are in your textbooks, whose discoveries power your phone, your GPS, your hospital: most of them didn't start out exceptional.

They started out exactly like you.

George Washington Carver was born into slavery. Literally. He was kidnapped as a baby and ransomed for a horse. He was turned away from a college he'd been accepted to when they saw his face. Told, explicitly, that Black men had no place in science.

He died having invented over 300 products from a single plant (the peanut) while running one of the most influential agricultural research programmes in American history. Thomas Edison tried to hire him. Henry Ford tried to hire him. He turned them both down.

Here's the part nobody puts in the textbook: every single morning, George Washington Carver prayed in his lab before he started work. He credited God for every discovery. Every. Single. One.

Isaac Newton was sent home from Cambridge when the plague shut the university down. He was 23, stuck at his mother's farmhouse with nothing to do.

In the next 18 months (alone, no professors, no funding, no colleagues) he invented calculus, discovered gravity, and proved that white light contains every colour in the spectrum.

Eighteen months.

Gregor Mendel failed his teaching exams. Twice. His supervisors thought his garden experiments were a waste of time. He spent 8 years counting peas in a monastery garden, alone, with zero recognition.

He died unknown.

Sixteen years later, three scientists independently found his work and realised he had already solved genetics. Today, every DNA test on earth runs on Mendel's laws.

What did all three have in common?

Not IQ. Not perfect circumstances. Not powerful connections.

This book has 30 of them: 30 Christian scientists who changed the world. And every single one shared the same thing.

You'll find it in Chapter 1. But once you see it, you won't be able to unsee it, and you won't be able to look at yourself the same way again.

This book is for you if:

You've ever felt like you don't belong in science. You've ever failed at something you actually cared about. You've ever wondered what you're supposed to do with your life. You've ever thought faith and intelligence couldn't exist in the same room.

They can. They always could. This book is the proof.

A note to anyone reading over a teenager's shoulder:

Everything in this book is true. The struggles are real. The faith is real. The failures are real. And the breakthroughs are real.

If there's a young person in your life who is brilliant but doubting, curious but discouraged, searching for proof that their gifts and their faith belong together: this is the book you've been looking for.

HOW TO USE THIS BOOK

You don't have to read it in order.

Find the chapter that speaks to where you are right now.

Struggling with feeling like you don't belong? Chapter 1 is yours. Tired of failing? Go straight to Chapter 2. Lost your sense of why any of this matters? Chapter 3 will find you.

Each story takes about 5 minutes. Each one ends with a Bible verse, a reflection, a prayer, and one thing you can actually do. Not homework. More like a dare.

No pressure. No timeline. No shame in reading one chapter and putting it down for a week.

Read one story. See if it changes anything. Then decide if you want another.

That's the whole plan.

One last thing before you start.

Every scientist in this book began with the same moment: a question they couldn't shake, a problem that kept nagging at them, something broken or beautiful or confusing that they couldn't stop thinking about.

None of them set out to be famous. None of them planned to change the world.

They just couldn't stop asking.

That feeling you have, the curiosity, the restlessness, the sense that there's something you're supposed to figure out. That's not random.

That's a calling.

Your experiment starts now.

Disclaimer and Legal Notice

TABLE OF CONTENTS

CHAPTER 1

Identity: Fearfully and Wonderfully Made

"I praise you because I am fearfully and wonderfully made; your works are wonderful, I know that full well."
<u>- Psalm 139:14</u>

Have you ever walked into a room and felt like you didn't belong?

Like everyone else had a secret you hadn't been given. Like they were born ready and you were still figuring out if you even deserved a seat at the table.

Every scientist in this chapter felt that exact feeling. Every single one of them was told, in one way or another, that they were the wrong person for the job.

And every single one of them changed the world anyway.

This chapter isn't about being the smartest. It's about knowing who made you, and trusting that he doesn't make mistakes.

GEORGE
WASHINGTON
CARVER
(1864-1943)
THE MAN WHO
ASKED WHAT
A PEANUT
COULD BECOME
A TRUE STORY OF
INVENTION, PURPOSE,
AND IMPACT

Right now, there is probably a jar of peanut butter somewhere in your kitchen. Maybe some peanut oil. Possibly shampoo, paint, or plastics made from plant ingredients.

Over 300 everyday products trace back to one man who spent his life asking a single question: what can this humble plant become?

That man was George Washington Carver. And the world spent years trying to make sure he'd never get the chance to find out.

George Washington Carver was born into slavery in Missouri, right in the middle of the American Civil War. He was so small and sick as a baby that people weren't sure he'd survive the week.

He almost didn't make it out of his first year alive.

Before he could walk, night raiders on horseback kidnapped him and his mother from the farm where they lived. His owner sent a man to find them. The man came back with baby George wrapped in a blanket. His mother was gone. She was never found.

George was ransomed for a horse. A single horse.

He grew up in a world that had already made up its mind about him. Black men didn't belong in science. That was just the way things were. When he was accepted to a college in Kansas, he packed everything he owned, saved every penny he had, and made the long journey to get there.

He was turned away at the front door when they saw his face.

The letter of acceptance meant nothing.

He found another school. Then another. He kept going, not because it was easy, but because something inside him simply would not let him stop. Every single morning, before the sun came up, he walked into his laboratory, got on his knees, and prayed. He asked God to show him the secrets hidden inside ordinary plants. The peanut. The sweet potato. The pecan.

He believed, with everything in him, that God had buried answers inside creation. His job was just to ask the right questions.

The answers came flooding in. Over 300 of them.

Thomas Edison, the most famous inventor in America, heard what Carver was doing and offered him a salary so large it would have made him one of the highest-paid scientists in the country.

Carver said no.

Henry Ford came next with an equally jaw-dropping offer.

Carver said no again.

He stayed at Tuskegee University, teaching poor farmers in the American

South how to bring their exhausted soil back to life. How to feed their families. How to build something from almost nothing.

When people asked him how he kept discovering so much, he smiled and said: "I never have to grope for methods. The method is revealed at the moment I am inspired to create something new."

He didn't do it for fame. He didn't do it for money. He did it because a question had taken hold of him as a child and never let go.

"For you created my inmost being; you knit me together in my mother's womb." - Psalm 139:13

Reflection and Prayer

George Washington Carver was told who he wasn't before he ever had the chance to show who he was. Has anyone ever made you feel like you didn't belong somewhere? In a classroom, on a team, in a subject you love? What would it look like to walk through that door anyway?

Lord, remind me that you made me on purpose. That no rejection, no closed door, and no voice telling me I don't belong can undo what you've already written into who I am. Help me

Your Challenge

Grab a bag of peanuts. Set a 10-minute timer. Write down every single thing you think a peanut could possibly become: foods, drinks, products, materials, anything your imagination can stretch to. Carver found 300. How far can you go?

MARY ANNING

(1799-1847)

The Girl the
Cliffs Kept
Calling
Back

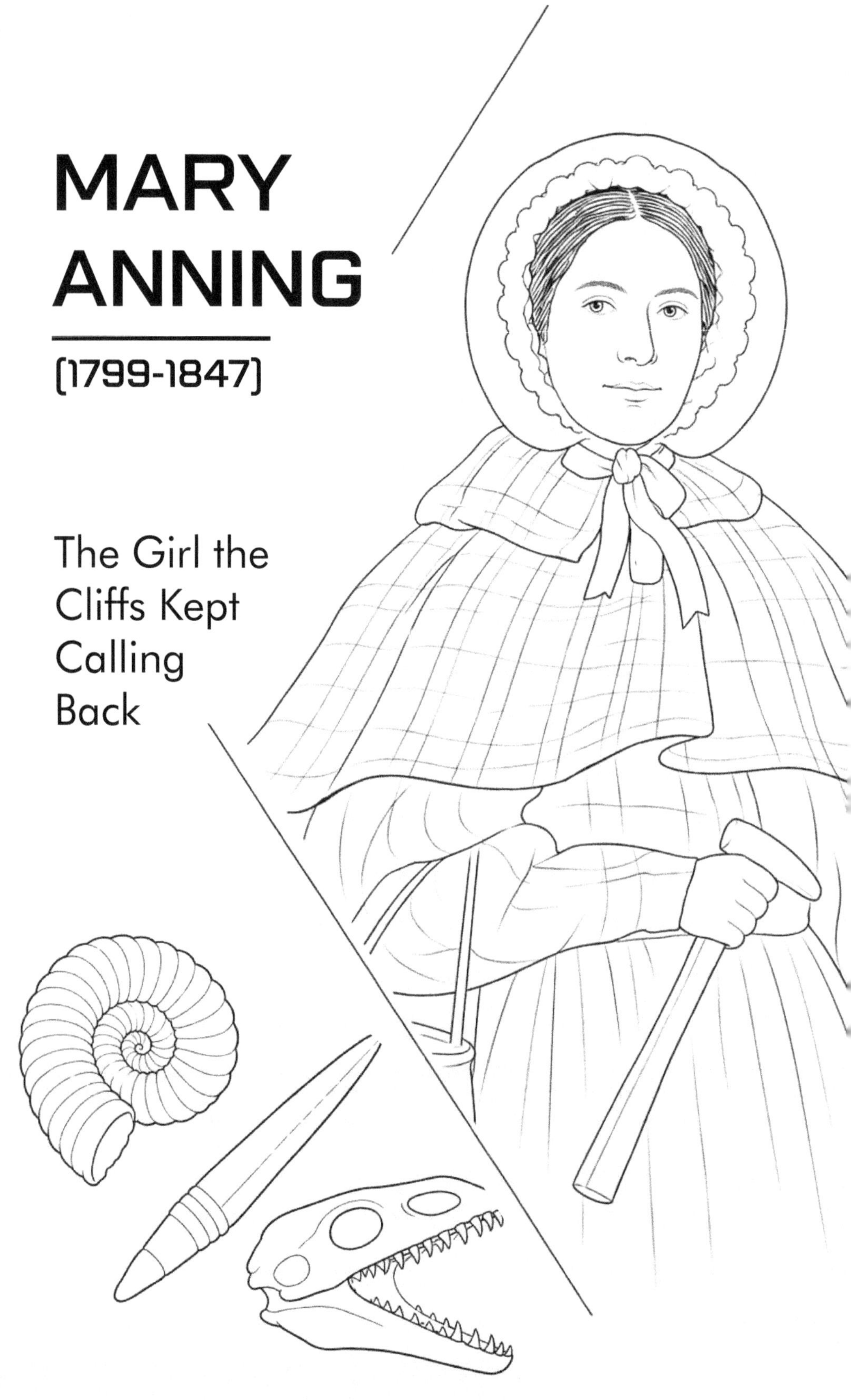

The next time you walk through a natural history museum and stare up at a prehistoric skeleton, here is something worth knowing: the science that put it there traces back to a girl who sold fossils on the beach to help her family buy food.

Her name was Mary Anning. She never earned a degree. She was never allowed into the scientific societies that argued over her discoveries. But she pulled creatures out of rock that no human being had ever seen before, and she changed how the world understands the history of life on earth.

Mary grew up in the seaside town of Lyme Regis, on the southern coast of England. The cliffs there were layered with fossils, strange shapes pressed into stone that her family sold to tourists for a few coins at a time. It was how they survived.

She started digging as a young girl. Not because anyone told her to. Because she genuinely could not stop thinking about what she kept finding.

What were these shapes? What had made them? How long ago had something living become stone?

When Mary was 12, she and her brother spotted something in the cliff face. A skull, half buried in rock, staring out at them from the grey stone. Mary spent months chipping away at the surrounding rock with careful, freezing hands, working through wind and rain on the exposed cliffside.

What finally came free was the first complete ichthyosaur (say it: ick-thee-oh-sore) ever found. An ichthyosaur was a dolphin-shaped sea reptile the size of a boat that lived 200 million years before Mary was born. Nobody alive had ever seen one. Nobody had known such creatures existed.

She was 12 years old.

She kept digging. She found a plesiosaur next: a long-necked sea creature with flippers like paddles and a neck almost as long as its body. Then a pterosaur, a flying reptile with wings like a bat, the first ever discovered in Britain.

Each find was extraordinary. Each one rewrote what the world understood about the creatures that had lived before us.

But the world Mary lived in had a problem. Credit didn't go to whoever made the discovery. It went to whoever had the right name, the right connections, and the right gender. Wealthy gentlemen who called themselves naturalists would travel to Lyme Regis, buy her finds for whatever she'd accept, and then present them to learned societies in London with their own names in the report.

The papers were written by men who had never touched the cliff face.

Mary never stopped digging.

Her faith held her steady through all of it. She saw the cliffs not as a challenge to God but as part of his creation: ancient, strange, and full of things he had made long before any human eye existed to find them.

When she died of cancer at 47, the Geological Society of London (which had never once admitted her as a member) stopped their meeting to read a tribute in her honour. It was the first time in their history they had ever done such a thing for anyone.

The questions Mary asked from a cold clifftop are still being answered today.

"I have called you by name; you are mine." - Isaiah 43:1

Reflection and Prayer

Mary's work was taken from her again and again. And again and again, she went back to the cliffs. Have you ever worked hard on something and watched someone else get the credit? How did that feel? What kept you going, or what made you stop?

God, when the world overlooks me, remind me that you see everything. You called me by name. My work has value even when no one else is watching. Give me the courage to keep digging. Amen.

Your Challenge

Go outside and find three rocks. Look at each one closely: really closely, like you're looking for something that no one else has noticed. Draw what you see on the surface of each one. Then ask yourself: how old might this be? What might it have been part of? You don't need a clifftop to start asking Mary's questions.

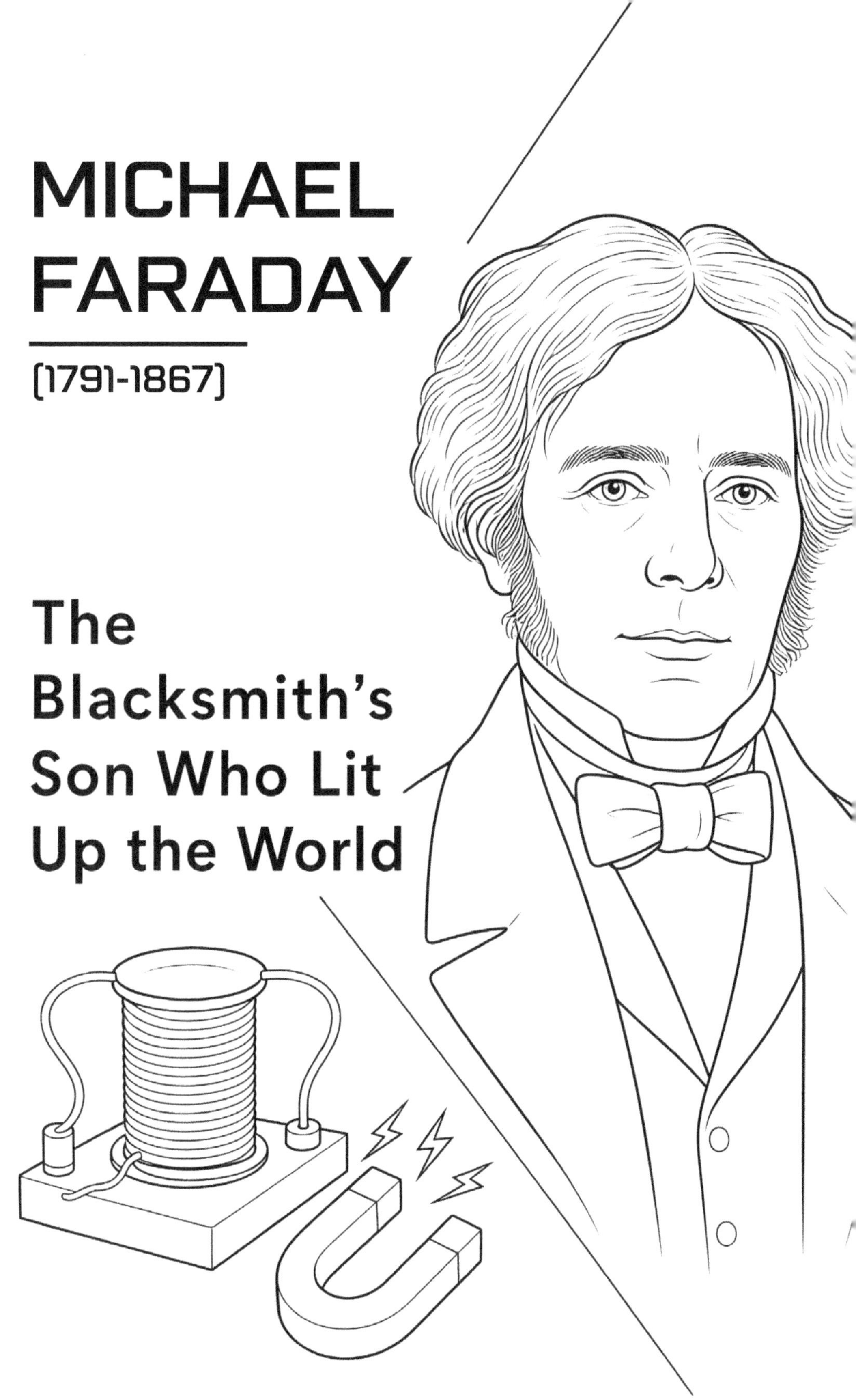

MICHAEL FARADAY
(1791-1867)
The Blacksmith's Son Who Lit Up the World

Every electric motor on earth runs on what Michael Faraday figured out. Every generator. Every electric car, every wind turbine, every power station keeping the lights on in your home right now. All of it traces back to a boy who grew up poor, never went to university, and couldn't afford a proper meal some nights.

He had no degree. No wealthy family. No one opening doors for him.

All he had was a question that refused to leave him alone. And it turned out to be enough.

Michael Faraday grew up in London in a cramped, cold apartment. His father was a blacksmith who was often too sick to work. There were weeks when the family shared a single loaf of bread between them and hoped it would last.

Michael left school at 13 to work as an errand boy for a bookbinder. He ran deliveries. He swept floors.

But books passed through that shop every single day. And Michael read every one he could get his hands on.

He read about electricity. About magnetism. About forces that were invisible and yet powerful enough to make metal move without touching it. He started building simple experiments in the back room after the shop closed. He attended free science lectures across the city, sitting quietly in the back row with his notebook open, writing down everything.

One day he heard that Humphry Davy, one of the most famous scientists in England, was giving a lecture nearby. Michael went. He took notes so detailed, so precise, so beautifully organised that he bound them into a little book and sent them to Davy with a letter asking for a job.

Davy wrote back. You can start Monday.

The scientists at the Royal Institution (Britain's most respected science organisation) looked down on Faraday for years. He had no degree. He came from nothing. Davy's own wife treated him like a servant. But Faraday kept his head down and kept experimenting.

His question: why does a wire twitch when you bring a magnet near it while electricity is flowing through it? Are electricity and magnetism actually connected? And if they are, what does that mean?

He spent 10 years on that question. He wrapped coils of wire around iron rings. He slid magnets in and out of tubes. He failed and failed and failed and came back the next morning and tried again.

His deep Christian faith gave him a kind of stubbornness that looked like peace. He believed God had designed the universe with order and purpose. That the rules were consistent. That if he kept looking carefully and honestly, the truth was there to be found.

It was.

What he discovered is called electromagnetic induction (which means: a moving magnet near a coil of wire creates electricity). That single discovery is the foundation underneath almost every source of electrical power on earth.

He was offered a knighthood. He turned it down. He was offered enormous sums of money. He turned those down too.

He said the work itself was the only reward that mattered.

"God chose the foolish things of the world to shame the wise; God chose the weak things of the world to shame the strong." - 1 Corinthians 1:27

Reflection and Prayer

Faraday was looked down on for years by people with better credentials and fancier backgrounds. Have you ever felt like your starting point was a disadvantage? What would it look like to let your curiosity be louder than anyone else's opinion of you?

Lord, thank you that you don't choose the way the world chooses. Help me to stop measuring myself by what I don't have and start using what I do have. Give me Faraday's patience: the kind that keeps showing up to the experiment even when no one believes in it yet. Amen.

Your Challenge

You can build what Faraday built. Search online for "how to make a simple electromagnet with a battery, wire, and a nail." Build it. Then hold it over a pile of paper clips and watch what happens. That discovery, the thing that powers your entire world, is right there in your hands.

Review

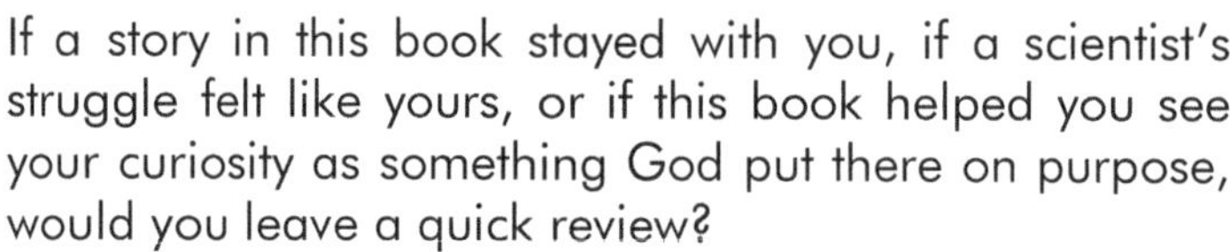

If a story in this book stayed with you, if a scientist's struggle felt like yours, or if this book helped you see your curiosity as something God put there on purpose, would you leave a quick review?

Even one sentence goes a long way. As a small author, your feedback helps another young scientist or thinker find this book when they need it most.

Scan the QR code below

with your phone camera to go straight to the review page.

Or go to your Amazon orders

find this book, and tap
"Write a product review."

**Thank you.
Genuinely.
It means everything.**

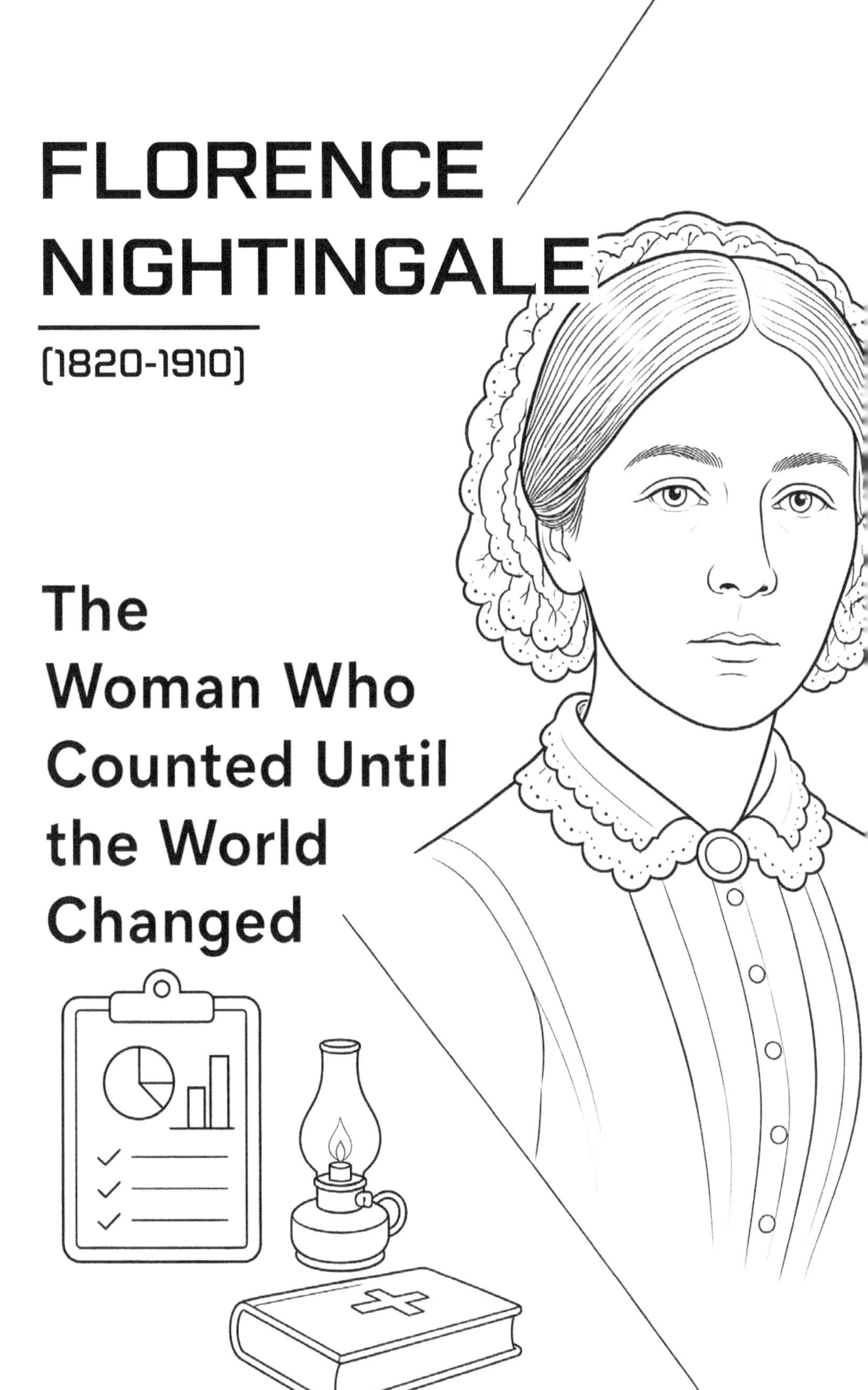

FLORENCE
NIGHTINGALE
(1820-1910)
The Woman Who Counted Until the World Changed

Every clean hospital. Every nurse who washes their hands before touching a patient. Every hygiene rule that keeps people alive after surgery. All of it exists because one woman refused to accept that so many people were dying for no good reason.

Her name was Florence Nightingale. She was told for 17 years that the work she felt called to do was not for someone like her. She did it anyway.

Florence was born into one of the wealthiest families in England in 1820. She grew up in beautiful country houses, surrounded by servants, educated by private tutors, dressed in the finest clothes.

The world had one future planned for her: marry a wealthy man, run a beautiful home, and live comfortably for the rest of her life.

Florence felt something pulling at her, something that would not leave her alone.

At just 16 years old, sitting quietly in the garden of her family's estate, she felt what she later described as a direct call from God. She didn't know yet what she was called to do. But she knew it was something specific. Something that mattered.

She started paying attention to sick people whenever she could. She visited the ill in the villages near her family's properties. She asked questions. She watched what helped and what didn't. She took notes in tiny handwriting and hid them away.

When she finally said aloud that she wanted to train as a nurse, her family was horrified. Nursing in 1837 was done by women no respectable family would speak to. It was dirty, exhausting, and considered far beneath someone of Florence's social class.

Her mother said no. Her father said no. She was forbidden.

She waited. She kept reading. She secretly built up a knowledge of hospitals and medicine that would have matched any doctor. For 17 years she was blocked at every turn, but she never stopped preparing.

Then came the Crimean War.

British soldiers were fighting in Turkey and dying in horrifying numbers. Not just from battle wounds, but from infections, cholera, and diseases that spread through the filthy, overcrowded military hospitals. A newspaper published the truth of what was happening, and the public was outraged.

Florence volunteered immediately. She arrived in the military hospital at Scutari to find overflowing wards, open sewers running beneath the floors, rats in the hallways, and patients lying in their own filth without clean water or proper food.

She got to work.

She organised the kitchens. She scrubbed the wards. She demanded proper drains. She sat up through the night with dying men, holding a lamp so they wouldn't face the darkness alone. The soldiers called her "the Lady with the Lamp."

But Florence's real weapon wasn't kindness. It was numbers.

She counted every death. She recorded every cause. She tracked every ward. Then she did something no one in medicine had ever done: she turned all that data into a picture. A circular diagram that made the pattern obvious at a single glance. Far more soldiers were dying from preventable infections than from enemy bullets.

She sent it to the British government.

They could not ignore what they were seeing. Hospitals were reformed. Standards were overhauled. The death rate in military hospitals fell from 42% to just 2%.

Florence Nightingale had gone to the Crimea to care for soldiers. She came back having used mathematics to change how governments protect public health forever.

She believed, until her final days, that every bit of it had been the answer to a calling she'd first heard as a 16-year-old girl sitting in a garden.

Act justly, love mercy, and walk humbly with your God." - Micah 6:8

Reflection and Prayer

Reflection and Prayer Florence waited 17 years. Seventeen years of being told no, of preparing in secret, of trusting that the door would open eventually. Is there something you feel called to that the people around you don't take seriously yet? What would it look like to keep preparing anyway?

God, give me Florence's patience. Help me to keep learning and preparing even when the doors stay shut. And when they finally open, give me the courage to walk through, even if what's on the other side is harder than I expected. Amen.

Your Challenge

Your Challenge The next time you are at a doctor's office, a hospital, or even just a pharmacy, write down five things you can see that are designed to keep you safe: sealed packaging, hand sanitiser, sterile gloves, clean surfaces, anything. Florence Nightingale fought her whole life to make those things standard. Before her, they didn't exist.

GREGOR
MENDEL

(1822-1884)

The Monk
Who Cracked
the Code of
Life

A A a
A

Every time a doctor says a disease "runs in families," they are using something Gregor Mendel figured out in a monastery garden more than 150 years ago.

Every DNA test. Every genetic diagnosis. Every conversation about inherited traits. All of it runs on laws that a monk discovered while counting peas, laws that the world ignored completely until 16 years after he died.

Gregor Mendel grew up in a small farming village in what is now the Czech Republic, the son of a peasant farmer who worked someone else's land for a living. Gregor was brilliant. His teachers could see it. But his family had almost no money, and in the 1840s, that meant your chances of getting a real education were close to zero.

He became an Augustinian friar (a member of a Christian religious community) partly because the monastery in Brno would pay for his studies.

Then, at 29, he failed his teaching exams.

Not just once. Twice.

The examiners said he didn't know enough. It was humiliating. The monastery sent him to the University of Vienna to study further. He came back two years later with a deep love for botany (the study of plants) and an obsession with mathematics and statistics.

Then he went into the garden and started growing peas.

He chose seven simple traits to study: plant height, pod shape, pod colour, seed shape, seed colour, flower colour, and flower position. He grew thousands of plants. He crossed them with each other in every combination he could think of. He kept meticulous records in small, neat handwriting, tracking every plant, every cross, every result.

It took eight years.

What he found was extraordinary. Traits didn't mix randomly. They followed precise mathematical rules. Some traits were dominant (they showed up in offspring no matter what). Others were recessive (they hid for a generation and then reappeared, like a message written in invisible ink). The ratios were consistent, predictable, repeatable.

He had cracked the code of inheritance.

He published his findings in a small regional science journal in 1866. He sent copies to scientists across Europe, including the most famous botanist in the world.

Most of them never wrote back.

The one who did said only that he had received the paper.

Mendel's work sat in libraries, untouched and uncited, for 35 years. He was eventually made the head of his monastery, buried in administrative work, and faded from the scientific world entirely. Near the end of his life he told a close friend: "My scientific work has brought me a great deal of satisfaction, and I am convinced that it will be appreciated before long by the whole world."

He died in 1884. Nobody in the scientific world knew his name.

In 1900, three scientists in three different countries, working completely independently, each made the same discovery about inherited traits. Each one searched the scientific literature to make sure they were first.

Each one found Mendel's paper.

He had already solved it. Thirty-five years earlier, alone in a monastery garden, he had already found the answer.

His patience outlasted the world's indifference.

**"Let us not become weary in doing good, for at the proper time we will reap a harvest if we do not give up."
- Galatians 6:9**

Reflection and Prayer

Mendel did eight years of careful, brilliant work that no one cared about in his lifetime. Have you ever worked hard on something and felt completely invisible? What does it feel like to keep going when no one seems to notice?

Lord, give me Mendel's faithfulness. Help me to do the work in front of me with care and precision, not for recognition, but because the work itself matters. And remind me that your timing is not always my timing. The harvest comes. Amen.

Your Challenge

Stand in front of a mirror. Pick three features: your eye colour, the shape of your nose, your earlobes (are they attached to your cheek, or do they hang free?). Ask a parent or grandparent which side of the family each one came from. You are reading Mendel's laws written on your own face. The same rules he found in 28,000 pea plants are the rules that made you.

CHAPTER 2

Resilience: 10,000 Ways That Won't Work

"I have not failed. I've just found 10,000 ways that won't work"
Thomas Edison

Here is the truth nobody puts on a motivational poster.

Every single scientist in this chapter failed. Publicly. Repeatedly. Some of them were laughed at. Some were taken to court. One was put under house arrest and forbidden from ever publishing again.

They kept going anyway.

Resilience is not a personality type. It is not something you're born with or you're not. It is a decision you make, again and again, every time the experiment blows up in your face. Every time the results don't match your prediction. Every time the person you most respect tells you that you're wrong.

The scientists in this chapter made that decision so many times it became habit. And their failures? They turned out to be the most important part of the story.

LOUIS PASTEUR
(1822-1895)
The Man Who Refused to Let Disease Win
VACCINE

The next time you drink a glass of milk, here is something worth knowing. That milk was heated to a specific temperature for a specific amount of time before it reached you. That process kills the dangerous bacteria that used to make milk deadly.

It is called pasteurisation. And it exists because a French chemist couldn't stop wondering why some wine batches went bad while others stayed perfectly fine.

Louis Pasteur grew up in a small tanning town in eastern France, the son of a leather worker. He was not a child prodigy. His early school reports described him as mediocre. But he was curious in a way that other students weren't, the kind of curious that doesn't let a question go until it has an answer.

He became a chemist. And in the 1850s, a local brewery came to him with a problem.

Some of their batches of beer were souring. Not all of them. Just some. And no one could figure out why.

Pasteur looked through his microscope and saw something unexpected: tiny living organisms were multiplying inside the liquid. In the good batches, one kind of organism. In the bad batches, a different kind entirely.

He had a wild idea. What if living microorganisms (microscopic creatures too small to see with the naked eye) were the cause of fermentation and spoiling? What if disease itself worked the same way?

The scientific establishment told him he was wrong. Disease, they said, was caused by bad air or by the body simply breaking down. The idea that invisible living creatures were spreading through food and water and bodies was considered almost laughable.

Pasteur kept going.

He designed elegant experiments that were almost impossible to argue with. He boiled broth and sealed it in curved-neck flasks that let air in but kept dust out. Nothing grew in the sealed floth. When he broke the necks open, microorganisms appeared within days. He had proved that life didn't appear from nowhere: it was carried.

His ideas saved the wine industry. Then the beer industry. Then, as he pushed the idea into medicine, they began saving human lives by the millions.

But the cost was enormous.

In 1868, Pasteur suffered a severe stroke at 45 years old. The left side of his body was partially paralysed. He could no longer perform experiments with his own hands. He lost three of his five children to typhoid, the very disease he was working to defeat.

He kept going.

He developed vaccines for cholera in chickens and anthrax in livestock. Then, in 1885, came his most dramatic moment. A nine-year-old boy named Joseph Meister was brought to his laboratory, his body torn by the bites of a rabid dog. Rabies was almost always fatal. There was no treatment.

Pasteur had been testing a rabies vaccine on dogs. It had never been tried on a human being. He was not a medical doctor. If he treated the boy and the boy died, his career would be destroyed.

He treated the boy anyway.

Joseph Meister lived. And the world changed.

"We also glory in our sufferings, because we know that suffering produces perseverance; perseverance, character; and character, hope." - Romans 5:3-4

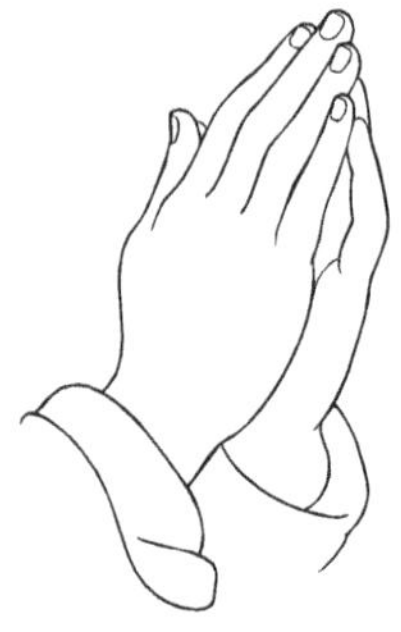

Reflection and Prayer

Pasteur lost his health, his mobility, and three of his children while doing the work he believed God had called him to. Have you ever kept going through something that cost you more than you expected? What gave you the strength to continue?

Lord, when the cost feels too high and the results feel too far away, remind me that perseverance builds something in me that success alone never could. Give me Pasteur's stubbornness. Give me his hope. Amen.

Your Challenge

Leave a small piece of bread in an open bag for three days. Check it each day and write down what you observe. What you're watching is exactly what Pasteur spent his life studying: microorganisms at work. Now ask yourself: where did they come from? How would you stop them?

JOSEPH
LISTER
(1822-1895)
The Doctor
Who Fought
the Enemy
Nobody
Could See
CARBOLIC
ACID

Every surgeon who washes their hands before an operation. Every nurse who wears sterile gloves. Every hospital ward that smells faintly of antiseptic. All of it exists because one doctor asked a question that his entire profession didn't want to hear.

Why do so many patients survive the operation but die in the days after?

Joseph Lister was born into a Quaker family in Essex, England. Quakers are a Christian community known for their deep commitment to honesty, simplicity, and caring for others. From his earliest years, Lister believed that his work as a doctor was a form of service to God.

He trained as a surgeon at a time when surgery was an almost desperate last resort. Not because operations were always dangerous. But because even when the operation itself went perfectly, a terrifying number of patients developed massive infections in the days afterward and died in agony.

Surgeons called it "ward fever." They accepted it as an unavoidable fact of medicine. The best hospitals in the world had death rates of 45% or higher after major operations.

Lister refused to accept it.

He read Louis Pasteur's work on microorganisms in 1865 and had an immediate, electric thought: what if those invisible organisms were getting into wounds and causing the infections? What if ward fever wasn't inevitable at all? What if it could be stopped?

He began experimenting with carbolic acid, a strong chemical used to treat sewage, as a way to kill microorganisms in wounds. He applied it to surgical instruments. He sprayed it into the air during operations. He soaked bandages in it.

The results were remarkable. His patients stopped dying of ward fever.

He published his findings and waited for the medical world to embrace them.

Instead, they pushed back.

Leading surgeons called his methods unnecessary and his theory unproven. Prestigious medical journals mocked the idea. Some of his colleagues simply refused to believe that invisible creatures they couldn't see were killing their patients.

Lister was calm, methodical, and relentless. He kept detailed records. He published more data. He demonstrated his methods again and again at hospitals across Europe.

Slowly, the evidence became impossible to ignore.

By the time Lister died at 84, antiseptic surgery had become standard practice everywhere in the world. The death rate from post-operative infection had collapsed from nearly half to almost nothing. It is estimated that his work has saved hundreds of millions of lives.

He was knighted by Queen Victoria. He was given honorary degrees by universities around the world. He was celebrated as one of the greatest benefactors in the history of medicine.

He said, simply, that he had tried to do what was right with what he had been given.

"Trust in the Lord with all your heart and lean not on your own understanding; in all your ways submit to him, and he will make your paths straight." - Proverbs 3:5-6

Reflection and Prayer

Lister watched patients die from something he believed he could stop, and then had to fight for years to convince a disbelieving world. Have you ever been certain about something that the people around you refused to accept? How did you handle the gap between what you knew and what others were willing to believe?

God, give me the courage to stand by what I know is true, even when the room disagrees. Give me Lister's patience: to keep showing the evidence, to keep doing the work, and to trust that the truth will eventually speak for itself. Amen.

Your Challenge

Pour a small drop of food colouring into a glass of water. Watch it spread. That is how Lister imagined microorganisms spreading through a wound, invisible, moving through the body's fluids. Now ask: what would stop it? What would you use to block the spread before it began?

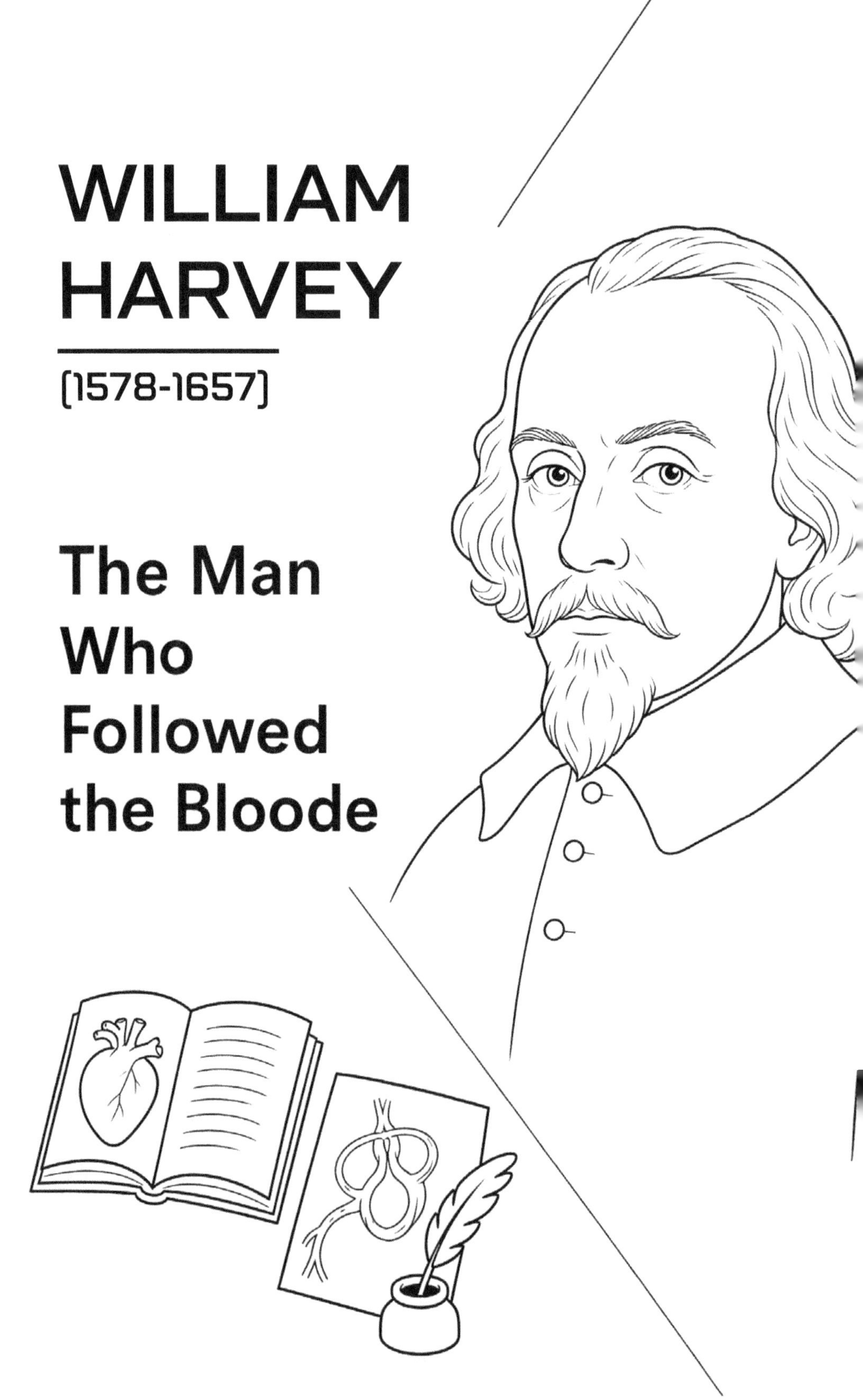

WILLIAM HARVEY
(1578-1657)

The Man
Who
Followed
the Bloode

Your heart is beating right now. It has been beating every second of every day since before you were born, around 100,000 times every 24 hours, pushing blood through 60,000 miles of blood vessels inside your body.

We know all of this because of William Harvey. A 17th-century English doctor who couldn't stop asking where the blood actually went.

William Harvey was born in Folkestone, England, in 1578. He trained as a physician in Cambridge and then in Padua, Italy, at the finest medical school in the world. He was sharp, careful, and unusually systematic in the way he approached problems.

The accepted wisdom about blood, handed down from the ancient Greek physician Galen 1,400 years earlier, went like this: the liver constantly produces new blood, the blood flows outward through the body, and the body absorbs it. New blood was always being made. Old blood was always being used up.

Harvey looked at this explanation and felt something nagging at him.

If the liver was constantly producing new blood, how much would that be? He did the maths. Based on how much blood the heart pumps with each beat, the body would need to produce hundreds of pounds of new blood every hour to make Galen's model work. That was obviously impossible.

So where did the blood actually go?

He began dissecting animals, dozens of different species, watching their hearts beat under his careful hands. He noticed that the valves in the heart only opened in one direction. Blood could not flow backward through them. It could only move forward.

He tied off veins and arteries in living animals and watched what happened. When he blocked a vein, the vessel swelled on the side away from the heart. When he blocked an artery, it swelled on the side toward the heart.

The blood was moving in a circle.

It left the heart through the arteries, travelled through the body, and returned through the veins. The same blood, over and over again, driven by the constant pumping of the heart.

He published his discovery in 1628 in a small, carefully worded book. The medical establishment responded with fury. Galen had been the authority on medicine for 14 centuries. Harvey was calling him wrong.

Colleagues stopped referring patients to him. His practice nearly collapsed. He was ridiculed at conferences and dismissed in letters.

He did not argue or fight back. He simply said: look at the evidence. Perform the experiments yourself. See what you find.

Twenty years after his publication, the circulation of blood had become universally accepted. Today, every single thing we know about the human heart, about blood pressure, about cardiovascular medicine, about surgery, rests on what Harvey discovered.

"For you created my inmost being; you knit me together in my mother's womb. I praise you because I am fearfully and wonderfully made." - Psalm 139:13-14

Reflection and Prayer

Harvey spent 20 years waiting for the world to catch up with what he already knew. Have you ever had to wait for people to recognise something that felt obvious to you? What does patience look like when you know you're right but the world isn't ready to agree?

Lord, thank you for the complexity of the body you designed. Give me Harvey's quiet confidence: the kind that doesn't need to shout, that trusts the evidence to speak, and that can wait calmly for the truth to be seen. Amen.

Your Challenge

Press two fingers gently on the inside of your wrist, just below your thumb. Feel your pulse. Count the beats for exactly 60 seconds. Write the number down. That is your resting heart rate. Now do 20 jumping jacks and count again. What changed? Harvey was the first person in history to understand the machine you just felt working.

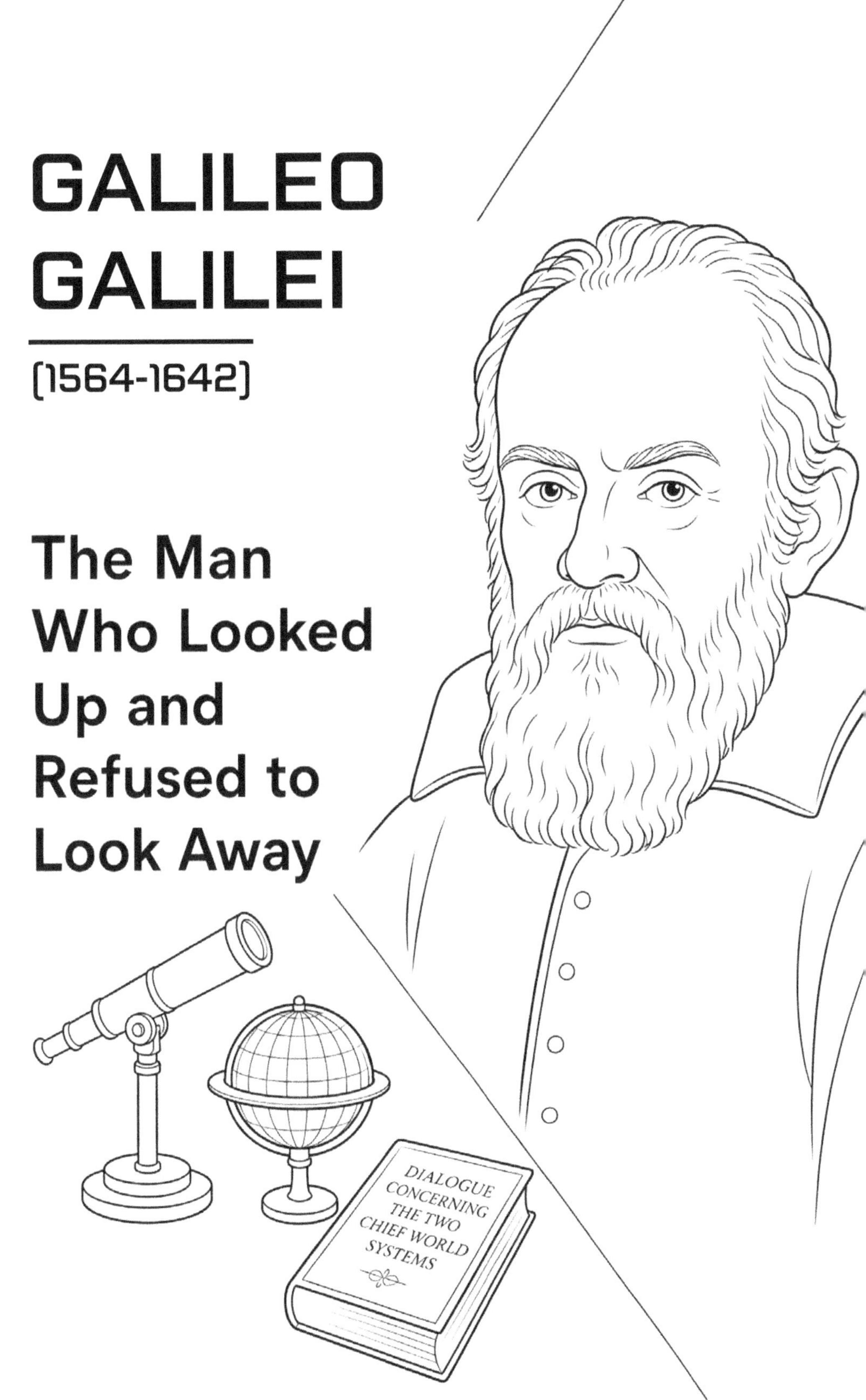

GALILEO GALILEI
(1564-1642)
The Man Who Looked Up and Refused to Look Away
DIALOGUE CONCERNING THE TWO CHIEF WORLD SYSTEMS

Every telescope ever built. Every image ever captured of a distant planet or a dying star. Every rocket ever launched into space. All of it begins with a man who sat in a cathedral one quiet afternoon, watching a chandelier swing, and couldn't stop thinking about what he was seeing.

His name was Galileo Galilei. And his curiosity eventually landed him in one of the most dramatic confrontations in the history of science.

Galileo was born in Pisa, Italy, in 1564, the son of a musician and music theorist. He was quick-witted, argumentative, and almost compulsively observant. As a young medical student, he noticed something strange about a bronze chandelier swinging in the cathedral of Pisa.

No matter how wide or how narrow the swing, the chandelier seemed to take the same amount of time to complete each arc.

He timed it with his own pulse. He was right.

He dropped out of medical school to study mathematics. He built inclined planes (ramps) and rolled balls down them at different angles, timing them obsessively. He discovered the basic laws of how objects fall and accelerate. He wrote detailed notebooks that fill entire libraries today.

Then he heard about a new invention from the Netherlands: a tube with lenses that could make distant things appear closer. Galileo built his own version within a day of hearing the description. Then he pointed it at the sky.

What he saw changed everything.

The moon was not a smooth, perfect sphere. It was covered in mountains and craters. Jupiter had four moons of its own orbiting around it. Venus went through phases like the moon. The sun had dark spots on its surface.

These were not abstract observations. They were devastating to the accepted picture of the universe, the one taught by the Church and inherited from ancient Greek philosophy, in which the earth stood motionless at the centre of everything and the heavens were perfect and unchanging.

Galileo became convinced that Copernicus had been right: the earth and planets moved around the sun.

He published his arguments. He was brilliant, witty, and thoroughly persuasive. He was also, some felt, needlessly aggressive in how he made his case.

In 1633, the Inquisition (the Church's powerful enforcement body) summoned him to Rome. He was 70 years old, ill, and barely able to travel. He was tried for heresy, the crime of teaching ideas contrary to Church doctrine.

He was found guilty.

He was forced to kneel before the court and publicly deny everything he had spent his life proving. He was sentenced to house arrest for the remaining years of his life, forbidden from publishing, forbidden from discussing his ideas.

He went home. He went blind. And he kept working.

He dictated his greatest scientific work, a comprehensive summary of everything he had discovered about motion and mechanics, to students who visited him in secret. It was smuggled out of Italy and published in the Netherlands in 1638.

Galileo died in 1642, still under house arrest, still a Catholic, still believing that the God who made the universe had also made the human mind capable of understanding it.

He never stopped looking up.

"Be strong and courageous. Do not be afraid; do not be discouraged, for the Lord your God will be with you wherever you go." - Joshua 1:9

Reflection and Prayer

Galileo was silenced, humiliated, and imprisoned for telling the truth. Have you ever been pressured to agree with something you knew was wrong? What does it cost to stay honest when the cost is high?

Lord, give me Galileo's courage: the kind that keeps working even when the doors are locked and the world says stop. Help me to trust that truth cannot be permanently buried, and that you are with me in the places where honesty feels dangerous. Amen.

Your Challenge

Find something that swings freely: a necklace, a keychain, or a piece of string with a weight tied to the end. Pull it back and let it swing. Time how long one full swing takes. Now change how far back you pull it. Does the time change? Galileo noticed this in a cathedral when he was supposed to be paying attention to the sermon. That observation eventually helped humanity build accurate clocks and navigate the seas.

CHAPTER 3

Purpose: Think God's Thoughts After Him

"I was merely thinking God's thoughts after Him." - Johannes Kepler

Why do you do what you do?

Most people never ask that question. They study because they are supposed to. They work because they have to. They get through the week without ever stopping to wonder what any of it is actually for.

The scientists in this chapter asked. And the answers they found transformed not just their work, but everything.

Every one of them believed that exploring the created world was a form of worship. That asking "how does this work?" was the same as asking "what did God make?" They weren't chasing fame or money or even the satisfaction of being right. They were chasing something that felt almost sacred: the chance to understand.

That kind of purpose doesn't run out. It doesn't burn out. It doesn't quit when things get hard.
It just keeps going.

JOHANNES KEPLER
(1571-1630)
The Man Who Read the Stars Like Scripture

Every satellite in orbit above your head right now is exactly where engineers calculated it would be. Every space mission that has ever reached another planet arrived because scientists could predict, with perfect accuracy, where that planet would be years into the future.

That precision exists because Johannes Kepler spent years staring at the sky and asking one relentless question: is there a hidden pattern in the way the planets move?

Johannes Kepler was born in 1571 in a small German town into a poor and troubled family. His father was unreliable, often away, possibly a mercenary soldier. His mother was later tried as a witch, a process Kepler spent years trying to stop. His childhood was not easy or safe.

But from his earliest years, the sky fascinated him.

He studied theology with the intention of becoming a Lutheran minister. But mathematics kept pulling at him. When a teaching position in mathematics opened in a small Austrian town, he took it, telling himself it was just temporary until something in the church came along.

The temporary position lasted the rest of his life.

Kepler became the assistant to Tycho Brahe, a Danish astronomer who had spent decades making the most precise measurements of planetary positions ever recorded. When Brahe died suddenly in 1601, he left Kepler his data: a treasure trove of observations more accurate than anything else in existence.

Kepler spent the next eight years trying to make the planets fit a circular orbit. He was certain the heavens moved in perfect circles. The Greek philosophers had said so. The Church had endorsed it. Everyone agreed.

The data didn't.

No matter how he adjusted his calculations, the numbers for Mars refused to line up on a circle. He tried every variation he could imagine. He filled hundreds of pages with calculations, threw them away, and started again.

Finally, exhausted and nearly defeated, he tried an ellipse.

It fit perfectly.

The planets didn't move in circles. They moved in ellipses, oval paths with the sun sitting slightly off-centre. And once Kepler accepted that single, uncomfortable truth, everything else fell into place. He discovered that planets speed up as they get closer to the sun and slow down as they move away. He found a precise mathematical relationship between a planet's distance from the sun and the time it takes to complete one orbit.

These became known as Kepler's three laws of planetary motion. They are still used today in every space mission ever launched.

Kepler described his discoveries as feeling like a moment of revelation. He wrote that in finding the mathematical architecture of the solar system, he was "thinking God's thoughts after Him." He was not using the words as poetry. He meant them literally. He believed that God had designed the universe according to mathematical principles, and that human minds, made in God's image, could actually read those principles.

He died in 1630, having spent his life doing exactly what he believed he was made to do.

"The heavens declare the glory of God; the skies proclaim the work of his hands." - Psalm 19:1

Reflection and Prayer

Kepler's purpose was clear to him: explore creation because it reveals the creator. Do you have a sense of what you were made to do? Not necessarily a career, but a kind of question or problem that genuinely lights you up?

Lord, give me the clarity Kepler had: the sense that my curiosity is not accidental, that it points somewhere. Help me to follow the questions you've placed inside me and to trust that in exploring your creation, I am somehow getting closer to knowing you. Amen.

Your Challenge

Look up a photo of the solar system's orbits. Notice they are not circles; they are slightly stretched, like eggs. Kepler spent eight years refusing to accept that fact, and then finally did. Draw your own ellipse by placing two pins in a piece of paper about 10 centimetres apart, looping a piece of string around both, and tracing the shape with a pencil held inside the loop. That oval is the shape of every planet's path around the sun.

ISAAC
NEWTON
(1643-1727)
The Man
Who Read
the Mind
of God
PHILOSOPHIÆ
NATURALIS
PRINCIPIA
MATHEMATICA

Your phone knows exactly where you are right now because satellites 20,000 kilometres above the earth are calculating your position using signals travelling at the speed of light. Those satellites stay in exactly the right orbits because engineers used equations to put them there.

Those equations were written by Isaac Newton. A farmer's son from Lincolnshire who spent 18 months alone during a plague outbreak and quietly rewrote the foundations of science.

Isaac Newton was born prematurely on Christmas Day, 1643, in a small farmhouse in Woolsthorpe, England. He was so small that his mother said he could have fitted inside a quart mug. Doctors didn't expect him to survive the week.

His father had died before he was born. When Newton was three, his mother remarried and moved away, leaving him in the care of his grandmother. He grew up quiet, solitary, and deeply interior.

He was not a standout student. His schoolmaster once described him as "idle" and "inattentive." But when he focused on something, he focused completely.

He went to Cambridge in 1661. Then, in 1665, the plague swept through England and Cambridge closed. Newton went home to Woolsthorpe.

In the next 18 months, alone in his childhood farmhouse, he invented calculus (a branch of mathematics that lets you calculate things that are continuously changing), discovered that white light is actually made up of every colour of the rainbow, and worked out the mathematical law that governs how every object in the universe attracts every other object.

He later described an apple falling in the orchard. He asked himself: if gravity pulls the apple to the ground, how far does gravity reach? Does it reach the moon?

He calculated. It reached the moon. It reached everything.

What is extraordinary about Newton is not just what he discovered. It is why he believed it mattered. Newton wrote more pages about theology and the Bible than about science. He believed, with absolute conviction, that the universe was designed by God according to precise, discoverable laws, and that the human mind, made in God's image, was capable of reading those laws.

He described himself not as a genius but as a man who had been "playing on the seashore, diverting myself in now and then finding a smoother pebble or a prettier shell than ordinary, whilst the great ocean of truth lay all undiscovered before me."

He spent his final years studying biblical prophecy, convinced that understanding scripture and understanding physics were ultimately the same project: reading the mind of God.

He died at 84, having reshaped science so completely that the world he left behind was barely recognisable from the one he was born into.

"He is before all things, and in him all things hold together." - Colossians 1:17

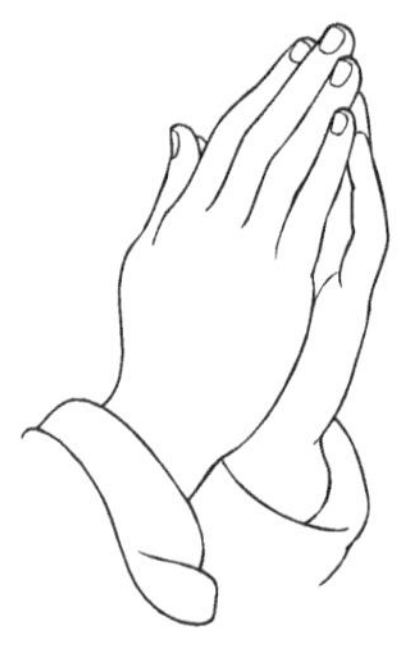

Reflection and Prayer

Newton saw no contradiction between faith and science. For him, they were the same search. Do you ever feel like your faith and your curiosity pull in different directions? What would it look like to let them work together?

Lord, thank you that you designed a universe with discoverable rules, and that you made minds capable of finding them. Help me to see curiosity as a form of worship, and to pursue understanding as a way of knowing you better. Amen.

Your Challenge

Drop two objects of very different weight from the same height at exactly the same time. A coin and a crumpled piece of paper work well. Which lands first? (The answer will surprise you if you haven't tried it.) Then look up why. Galileo tested this first; Newton explained it. You just replicated one of the most important experiments in the history of physics.

ROBERT
BOYLE
(1627-1691)
The Chemist
Who Gave
Everything
Away
THE
SCEPTICAL
CHYMIST

Every time a doctor checks your blood pressure, every scuba tank filled with compressed air, every car tyre inflated at a petrol station: all of it relies on a principle that Robert Boyle discovered when he couldn't stop wondering what air was actually made of.

He was one of the wealthiest men in England. He could have done anything.

He chose chemistry. And he gave almost everything he earned away.

Robert Boyle was born in 1627 into the family of the Earl of Cork, one of the richest men in Ireland. He had tutors, estates, and every advantage money could provide. He was educated at Eton. He travelled Europe as a young man, visiting the great minds of the age.

He could have lived a life of comfortable leisure. Instead, he set up a laboratory.

What fascinated him was matter itself. What is the world actually made of? What is air? What happens inside a liquid? What is fire? These questions were not idle curiosities for Boyle. He believed that God had hidden astonishing truths inside the physical world, and that it was both a privilege and a responsibility to find them.

He worked with a brilliant assistant named Robert Hooke to build a powerful air pump, one of the most sophisticated scientific instruments in the world at that time. Together they removed air from sealed chambers and watched what happened: flames went out, sounds became muffled, animals lost consciousness.

Air was not just nothing. It was something. And it had properties that could be measured.

Boyle went further. He trapped air in a sealed tube and compressed it, pushing a column of mercury down onto it with steadily increasing force. He found a precise, reliable relationship: as the pressure on the gas doubled, the volume of the gas halved. As pressure decreased, volume increased.

This became Boyle's Law. It is one of the first quantitative laws in the history of chemistry: a precise mathematical description of how matter behaves. It opened the door to modern chemistry.

But Boyle's faith drove everything he did beyond the laboratory. He taught himself Greek and Hebrew to read scripture in its original languages. He spent large portions of his personal fortune funding the translation of the Bible into languages spoken by people who had never had access to it: Irish, Welsh, Turkish, Malay. He set up a lecture series specifically to defend the Christian faith against atheism and deism, which continued for decades after his death.

He believed that chemistry and theology were not separate pursuits. Both were ways of understanding what God had made and what God had said.

He never married, never sought titles, and gave away wealth that would have made most people comfortable for generations.

When he died in 1691, he left most of what remained to charities and to the church.

"For since the creation of the world God's invisible qualities, his eternal power and divine nature, have been clearly seen, being understood from what has been made." - Romans 1:20

Reflection and Prayer

Boyle had every reason to coast through life on wealth and comfort. He chose something harder and more meaningful instead. Are there areas of your life where comfort is pulling you away from something you know matters more?

Lord, give me Boyle's generosity: the kind that doesn't cling to what it has been given, but pours it back out in service to others. Help me to use whatever I have been given, wealth, talent, time, in ways that point people toward you. Amen.

Your Challenge

Blow up a balloon halfway and tie it. Squeeze it as hard as you can without popping it. Notice how the balloon resists you. That resistance is Boyle's Law in action: you are increasing the pressure, so the gas pushes back harder to maintain its volume. Now hold the balloon in front of a heater or in cold air from a refrigerator. What happens? You've just discovered something Boyle spent years exploring.

FRANCIS COLLINS

(1950-)

The Atheist Who Read the Code and Found God

Every vaccine developed at record speed during a global pandemic. Every genetic test that tells a family whether a disease will be passed to their children. Every cancer treatment designed around a patient's specific DNA.

All of it accelerated because of a project led by a man who started his career as a committed atheist, read the language written inside every human cell, and completely changed his mind.

Francis Collins grew up on a small farm in the Shenandoah Valley of Virginia, the son of two artists who believed in education above everything else. He was homeschooled until sixth grade, taught to love learning for its own sake.

He had no strong religious beliefs. By the time he reached university and then medical school, he had arrived at what he considered a reasonable conclusion: science explained the world well enough that God wasn't necessary.

Then he went to work in a rural hospital in North Carolina.

He saw patients facing death with a serenity he couldn't explain. One elderly woman in particular, dying of a painful disease, asked him one afternoon what he believed. When he said he wasn't sure, she looked at him gently and said: "It's okay not to be sure. But have you actually thought about it?"

He hadn't. Not really.

He started reading. He worked through C.S. Lewis's book Mere Christianity, expecting to find it easily dismissible. He found it wasn't. The arguments were careful, honest, and harder to knock down than he expected.

Then he did what scientists do. He followed the evidence.

He became a Christian in his early thirties, kneeling alone on a hiking trail in the Cascade Mountains, looking at a frozen waterfall.

His faith didn't slow down his science. It intensified it.

In 1993, Collins was appointed to lead the Human Genome Project, the most ambitious scientific undertaking in history: a global effort to read and map every single piece of DNA in the human body. The human genome contains over 3 billion chemical letters, the complete instruction manual for building and running a human being.

It took 13 years and the work of thousands of scientists on multiple continents.

When the first draft was completed in 2000, Collins stood beside President Bill Clinton at the White House and described the human genome as "the language in which God created life."

He saw no contradiction. He saw the genome as one of the most stunning things he had ever encountered: a text of almost incomprehensible complexity and elegance, written into every cell of every human body.

He continued working in science and medicine for decades. During the COVID-19 pandemic, as director of the National Institutes of Health, he helped oversee the fastest vaccine development in human history.

He still prays every day. He still believes the two searches, for truth in scripture and for truth in creation, ultimately lead to the same place.

"In the beginning was the Word." - John 1:1

Reflection and Prayer

Collins was a scientist who followed the evidence all the way to God. Have you ever changed your mind about something important because the evidence demanded it? What does it look like to be genuinely open to being wrong?

Lord, give me Collins' honesty: the kind that follows evidence wherever it leads, even when the destination is unexpected. Help me to hold my faith and my curiosity together, trusting that you are the author of both. Amen.

Your Challenge

Search online for "what is DNA" and watch a short explainer video, ideally under 3 minutes. Then consider this: every single cell in your body contains roughly 2 metres of DNA, coiled up so tightly it fits inside a space far smaller than the tip of a pin. You have about 37 trillion cells. Francis Collins read that code and found God written into it. What do you find?

CHAPTER 4

Focus: Give Me a Lever Long Enough

"Give me a lever long enough and a fulcrum on which to place it, and I shall move the world." - <u>Archimedes</u>

The world will tell you to keep your options open.

Try everything. Explore every direction. Don't put too much energy into any one thing too soon. Stay flexible. Stay broad.

The scientists in this chapter ignored that advice completely.

Each of them found one question they could not let go of. Not a general topic. Not a vague interest. One specific, burning, nagging question that showed up in their mind first thing in the morning and last thing at night.

And they went deeper into it than anyone had ever gone before.

> A lever doesn't need to be big. It needs to be placed exactly right. One precisely aimed question, pursued with complete focus, can move things that no amount of scattered effort ever could.

JAMES CLERK MAXWELL

(1831-1879)

The Man Who Proved Everything is Connected

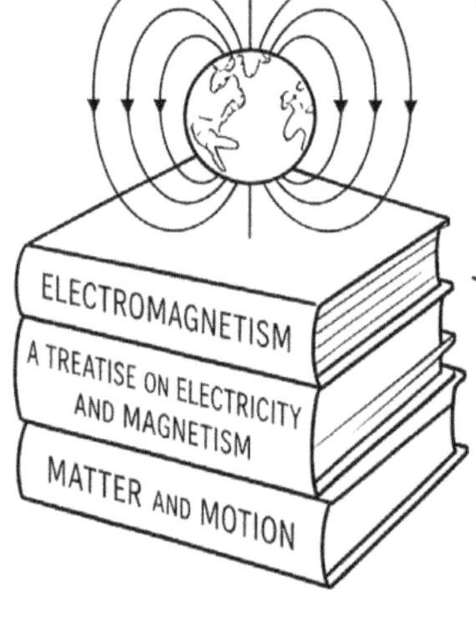

$$\nabla \cdot \mathbf{E} = \frac{\rho}{\varepsilon_0}$$

$$\nabla \cdot \mathbf{B} = 0$$

$$\nabla \times \mathbf{E} = -\frac{\partial \mathbf{B}}{\partial t}$$

$$\nabla \times \mathbf{B} = \mu_0 \mathbf{J} + \mu_0 \varepsilon_0 \frac{\partial \mathbf{E}}{\partial t}$$

Your Wi-Fi signal is an electromagnetic wave. So is the light coming from your lamp. So are the X-rays taken at the hospital, the radio waves carrying your favourite songs, and the invisible radiation keeping your food warm in the microwave.

They are all the same thing, at different frequencies.

James Clerk Maxwell was the first person to prove it, using four elegant equations that Einstein called the most profound scientific achievement since Newton.

James Clerk Maxwell was born in Edinburgh, Scotland, in 1831, the only child of a lawyer who was genuinely curious about how everything worked. His childhood home was a place where questions were welcomed, and Maxwell grew up treating the world as one long, fascinating puzzle.

He started writing scientific papers at 14. Not because a teacher told him to. Because he had a question and writing was how he worked through it.

The question that consumed him: what is light? Not as a philosophical matter, but as a physical one. It moved like a wave. It had speed. It behaved in predictable ways. But what was it actually a wave of?

At the same time, electricity and magnetism were being studied intensively across Europe, but separately. Electricity did this. Magnetism did that. No one had found the underlying connection.

Maxwell found it.

Working through years of dense mathematics, he developed a set of equations that described electricity and magnetism as two expressions of the same underlying force. And then he noticed something extraordinary in his own equations: they predicted the existence of a wave that travelled at the speed of light.

He recognised immediately what that meant.

Light was an electromagnetic wave.

His equations also predicted that other electromagnetic waves should exist, at different frequencies and energies. Radio waves. Microwaves. X-rays. All of them were implied by his mathematics, decades before anyone built a device to produce or detect them.

Heinrich Hertz confirmed the existence of radio waves experimentally in 1887, eight years after Maxwell died. Every technology built on radio, Wi-Fi, radar, satellite communications, was vindicated by what Maxwell had written at his desk in Scotland.

Maxwell was a deeply devout evangelical Presbyterian Christian. His faith gave him a settled, unhurried quality that those who knew him found unusual. He believed the universe was designed to be understood, that mathematics was the language God had used to write it, and that focused, patient inquiry was the most honest form of worship available to a scientist. He died of stomach cancer at 48. He had changed physics forever and barely made the news.

"It is the glory of God to conceal a matter; to search out a matter is the glory of kings." - Proverbs 25:2

Reflection and Prayer

Maxwell spent his career chasing one deep question, the connection between electricity, magnetism, and light, and found that the answer was more beautiful and far-reaching than he had imagined. Is there a question in your life that you keep returning to? What would it look like to go deeper into it instead of broader?

Lord, help me to resist the pressure to scatter my attention in every direction. Show me the question you've placed inside me, and give me the discipline and the courage to pursue it all the way to the bottom. Amen.

Your Challenge

Turn on a torch, your phone's screen, your Wi-Fi router, and a radio (or music from a Bluetooth speaker) in the same room. Every single one of those devices is producing electromagnetic waves. Maxwell's equations describe all of them. Can you find two more sources of electromagnetic waves in the same room? (Hint: the microwave oven and the TV remote both count.)

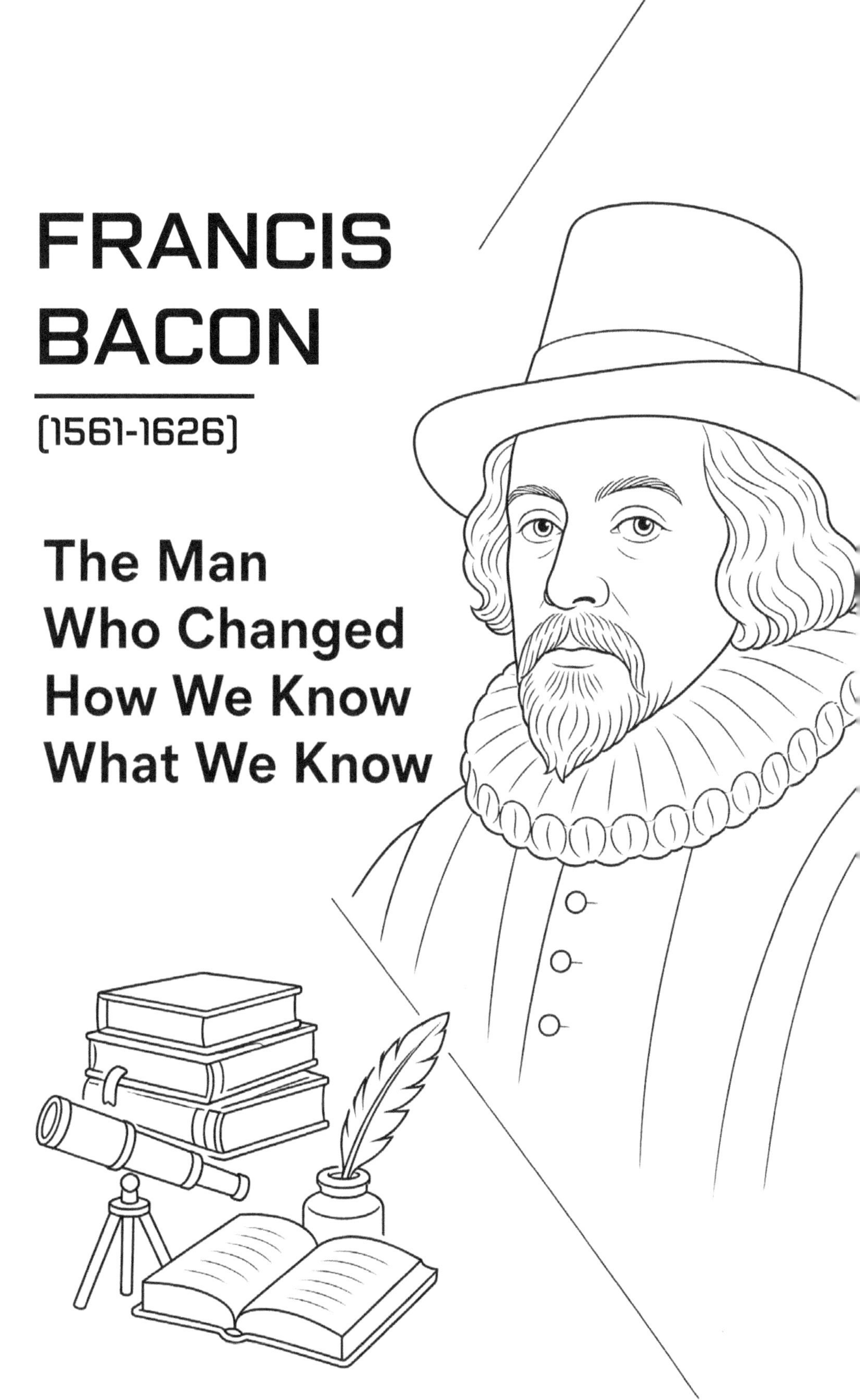

FRANCIS BACON
(1561-1626)
The Man
Who Changed
How We Know
What We Know

Every experiment ever performed in a school science lab. Every clinical drug trial that decides whether a new medicine is safe. Every research paper published in any scientific journal anywhere on earth.

All of it follows a process that Francis Bacon designed.

Not because he discovered something dramatic. But because he asked a question that no one before him had seriously confronted: how do we actually know when something is true?

Francis Bacon was born in London in 1561, the son of a senior royal official. He was educated at Cambridge from the age of 12 and trained as a lawyer. He rose to become one of the most senior legal figures in England, eventually serving as Lord Chancellor.

But what genuinely obsessed him was a problem he saw everywhere he looked.

Smart, educated, respected people were making confident claims about the natural world based on ancient authorities. Aristotle had said it. Galen had said it. The accepted wisdom was the accepted wisdom, and questioning it was seen as arrogance at best and heresy at worst.

But what if Aristotle was wrong? What if Galen was wrong? How would anyone know? No one was checking. No one was testing. No one was designing experiments to see whether the accepted claims held up against reality.

Bacon found this maddening.

He spent years developing what he called a new method for acquiring knowledge. The idea sounds simple today, but was radical in the early 1600s: don't start with what authorities have said. Start with what you can observe. Collect data. Look for patterns. Test your conclusions by trying to prove them wrong, not just right. Build knowledge slowly, carefully, from the ground up.

This is the scientific method. Every scientist who has ever lived since Bacon has used it, usually without knowing his name.

He was an Anglican Christian who believed that two things revealed God to humanity: scripture and creation. He called them the "book of God" and the "book of nature," and argued that studying nature carefully was a form of honouring the God who made it.

He didn't invent an instrument or discover a law. He invented something more foundational: the way we decide what to believe.

"Test everything. Hold on to what is good." - 1 Thessalonians 5:21

Reflection and Prayer

Bacon's contribution wasn't a discovery; it was a way of thinking. He asked: how do we make sure we're not fooling ourselves? Is there an area of your life, a belief, an assumption, a habit, where you have never actually tested whether it's true?

Lord, give me Bacon's honesty: the courage to question what I think I know, to look at the evidence rather than just the authority, and to follow truth wherever it leads even when it challenges what I assumed. Amen.

Your Challenge

Pick one thing you believe but have never actually tested. It can be small: "warm water freezes faster than cold water," or "I'm worse at maths in the morning." Design a simple experiment. Run it. Write down what you find. You just did exactly what Bacon spent his life trying to teach the world to do.

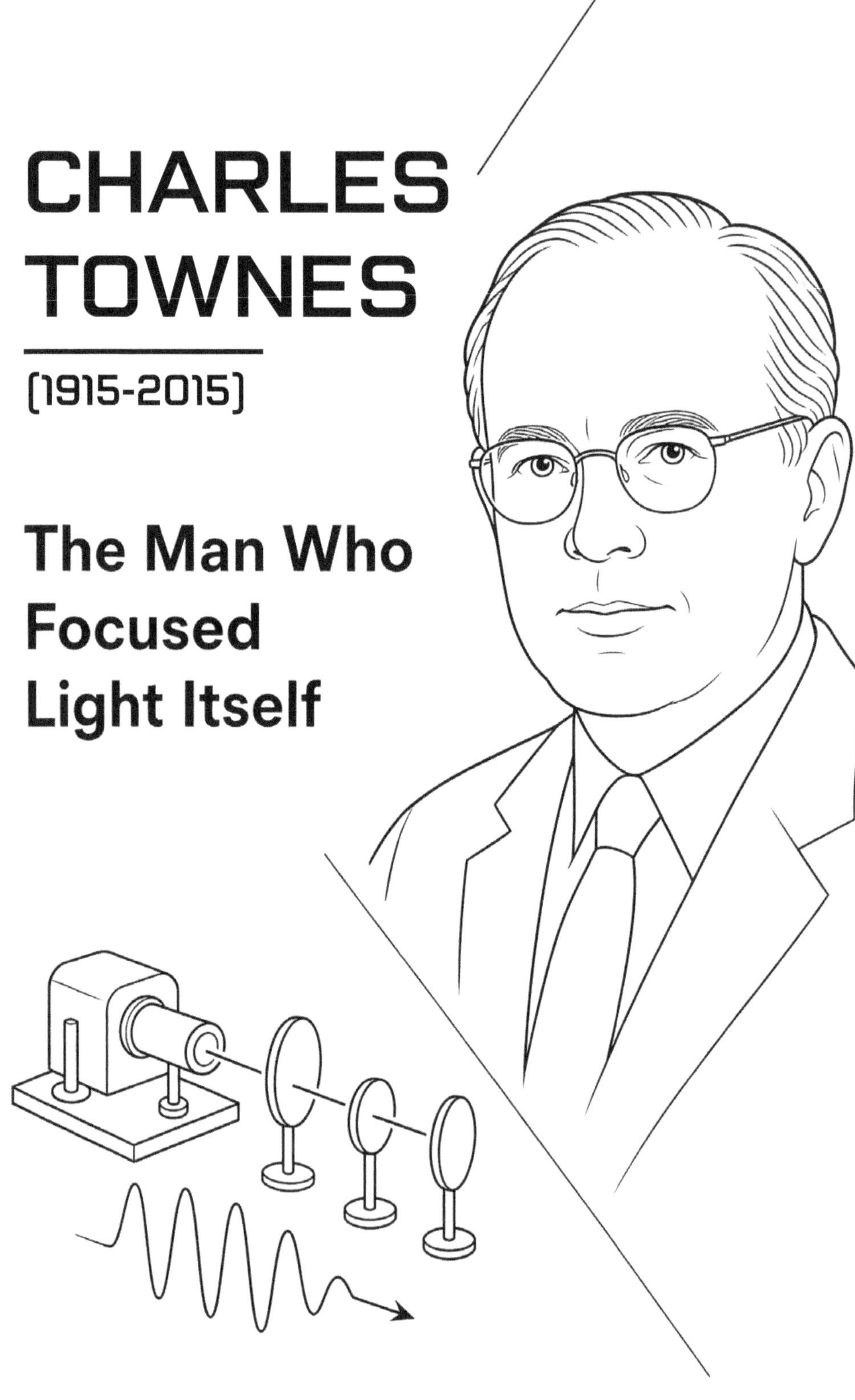

CHARLES TOWNES
(1915-2015)
The Man Who Focused Light Itself

Every barcode scanned at a shop checkout. Every laser eye surgery performed in a hospital. Every DVD ever played, every construction laser ever used to align a wall, every fibre optic cable carrying internet signals under the ocean.

All of it depends on the laser. A device that didn't exist until Charles Townes sat on a park bench one morning and had an idea so precise and specific that he wrote it on the back of an envelope before he stood up.

Charles Townes was born in Greenville, South Carolina, in 1915, into a Christian family who believed that curiosity was a gift and learning was a duty. He was fascinated by nature from childhood, collecting insects and plants and spending hours in the fields around his home simply looking at things.

He studied physics at university and became particularly interested in a question about molecules (the tiny clusters of atoms that make up everything around you): why did molecules absorb and emit energy at such strikingly precise and consistent frequencies?

This was not a glamorous question. It was not the sort of thing that made the news. But it gnawed at Townes. He suspected that the answer had practical implications, that the precise energy frequencies of molecules could be used to do something useful, though he wasn't yet sure what.

In 1951, he was in Washington D.C. for a scientific conference, sitting on a bench in a nearby park early in the morning. The azalea bushes were in full bloom around him, bright pink in the grey spring air. He had been wrestling with the problem for months.

The idea arrived all at once.

Molecules could be stimulated to release their energy in a controlled, synchronised burst. Instead of light scattering in every direction as it does from a normal bulb, you could force it to travel in a perfectly aligned beam, every wave in step with every other wave, all the same frequency, all the same direction.

He grabbed the envelope in his jacket pocket and scribbled equations.

What he was describing, without quite knowing the full implications yet, was the laser.

Senior colleagues told him it was impossible. They told him the idea violated fundAmen.tal physical laws. A Nobel Prize-winning physicist told him to his face that he was wasting his time.

Townes kept going.

He built a working prototype, first called a maser (using microwave frequencies), then extended the principle to visible light and called it a laser

(Light Amplification by Stimulated Emission of Radiation). In 1964, he was awarded the Nobel Prize in Physics.

Townes was a member of the United Church of Christ and prayed daily for his entire career. He wrote and spoke extensively about the relationship between science and faith, arguing that both were fundAmen.tally searches for truth and that the two were not in conflict.

He said science gave him a deep appreciation for the elegance of creation, and faith gave him the trust that the answers were worth looking for.

He was 99 years old when he died in 2015, having lived to see the laser transform medicine, communications, manufacturing, and science itself.

"And the peace of God, which transcends all understanding, will guard your hearts and your minds." - Philippians 4:7

Reflection and Prayer

Townes was told his idea was impossible by people who were supposed to know better. Have you ever been discouraged from pursuing something by someone whose opinion you respected? How did you decide whether to listen or to keep going?

Lord, give me Townes' clarity: the ability to focus completely on what you've put in front of me, to resist the noise of discouragement, and to trust that the idea you've placed in my mind is worth pursuing all the way. Amen.

Your Challenge

Shine a torch at a wall from across the room. Notice how the light spreads out and becomes dimmer as it travels. Now search online for a video showing a laser beam travelling the same distance. Notice that it barely spreads at all. That difference, scattered light versus perfectly focused light, is what Townes spent decades chasing. What could you do with light that never scattered?

LORD
KELVIN
(1824-1907)

The Man
Who Found
the True Zero

Every physics equation ever written uses temperature. And every one of those equations uses a scale that starts at the only temperature that makes absolute, undeniable sense: the point where all heat completely stops.

That point is called absolute zero. It is -273.15 degrees Celsius. It is 0 on the Kelvin scale.

William Thomson, who became Lord Kelvin, found it because he couldn't accept that every existing temperature scale was built on an arbitrary starting point chosen for human convenience rather than physical reality.

William Thomson was born in Belfast in 1824, the son of a mathematics professor. He was so gifted that he enrolled at Glasgow University at the age of ten. Not as a special student. As a regular student, attending the same classes as everyone else.

He published his first scientific paper at 16.

What fascinated him from early in his career was thermodynamics, the science of heat and energy, and specifically the question of temperature itself.

The Celsius scale set zero at the freezing point of water. Fahrenheit set zero at the temperature of a specific ice and salt mixture. Both scales were practical and useful for everyday life. But both of them were arbitrary. Both of them allowed temperatures below zero, which implied the possibility of an infinite regression downward.

Thomson kept asking: is there a true bottom? A point where heat genuinely, absolutely, finally stops?

He approached the problem through careful mathematical analysis of how gases behave as they cool. As temperature drops, gas pressure drops in a predictable, linear way. When Thomson extended that line mathematically, it pointed to a specific temperature at which pressure would reach zero: -273.15 degrees Celsius.

That had to be the floor. That had to be absolute zero. Because pressure cannot go below zero; there is no such thing as negative pressure in this physical sense. So there had to be a temperature below which nothing could go.

He proposed a new temperature scale starting at that point. Zero would mean zero. Not "zero for our purposes" but actually zero. No heat at all.

The Kelvin scale is now the international standard for scientific temperature measurement. Every equation in physics and chemistry that involves temperature uses it.

Kelvin was a devout Christian throughout his life. He wrote and spoke about the relationship between faith and science, argued strongly for the existence of a creator God, and saw his scientific work as a way of exploring and honouring the designed order of the universe.

He also led the project to lay the first transatlantic telegraph cable, connecting Britain and America for the first time, a project he pursued with the same methodical precision he brought to everything.

He was 83 when he died, still asking questions, still convinced that the universe had more to reveal.

"I am the Alpha and the Omega, the First and the Last, the Beginning and the End." - Revelation 22:13

Reflection and Prayer

Kelvin's great insight was to ask: is there an actual limit? A real bottom? A true starting point? Is there an area of your life where you are measuring things against an arbitrary standard rather than asking what the real foundation actually is?

Lord, you are the true starting point of everything. Help me to build my life on what is actually real rather than what is merely convenient. Give me Kelvin's precision: the desire to get things exactly right, not just approximately right. Amen.

Your Challenge

Look up the following temperatures and convert each one to Kelvin by adding 273.15 to the Celsius value: the surface of the sun, liquid nitrogen, the coldest temperature ever recorded on earth, and the temperature of deep space. Notice that in Kelvin, none of them are negative. That's the point. Kelvin's scale starts where heat actually starts: at nothing.

CHAPTER 5

Discipline:
The Compound Effect

"Compound interest is the eighth wonder of the world. He who understands it, earns it; he who doesn't, pays it." - Albert Einstein

Nobody saw these scientists working. That is the point.

The breakthrough moments, the discoveries, the publications that changed everything: those are the parts that get remembered. But they are the smallest part of the story. The real story is what happened before. The years of notebooks. The repeated experiments. The questions revisited after every wrong answer. The early mornings and the late nights when no one was watching and nothing seemed to be working.

Greatness is not a moment. It is a habit compounded over time.

One percent better every day for a year doesn't add up to 365 percent improvement. Because of compound interest, it adds up to 3,700 percent.

The scientists in this chapter understood this. Some of them spent decades on a single problem. One of them kept working after he went completely blind. They didn't achieve extraordinary things because they were extraordinary people.

They achieved extraordinary things because they showed up. Every single day.

LEONHARD EULER

(1707-1783)

The Man Who Never Stopped Thinking

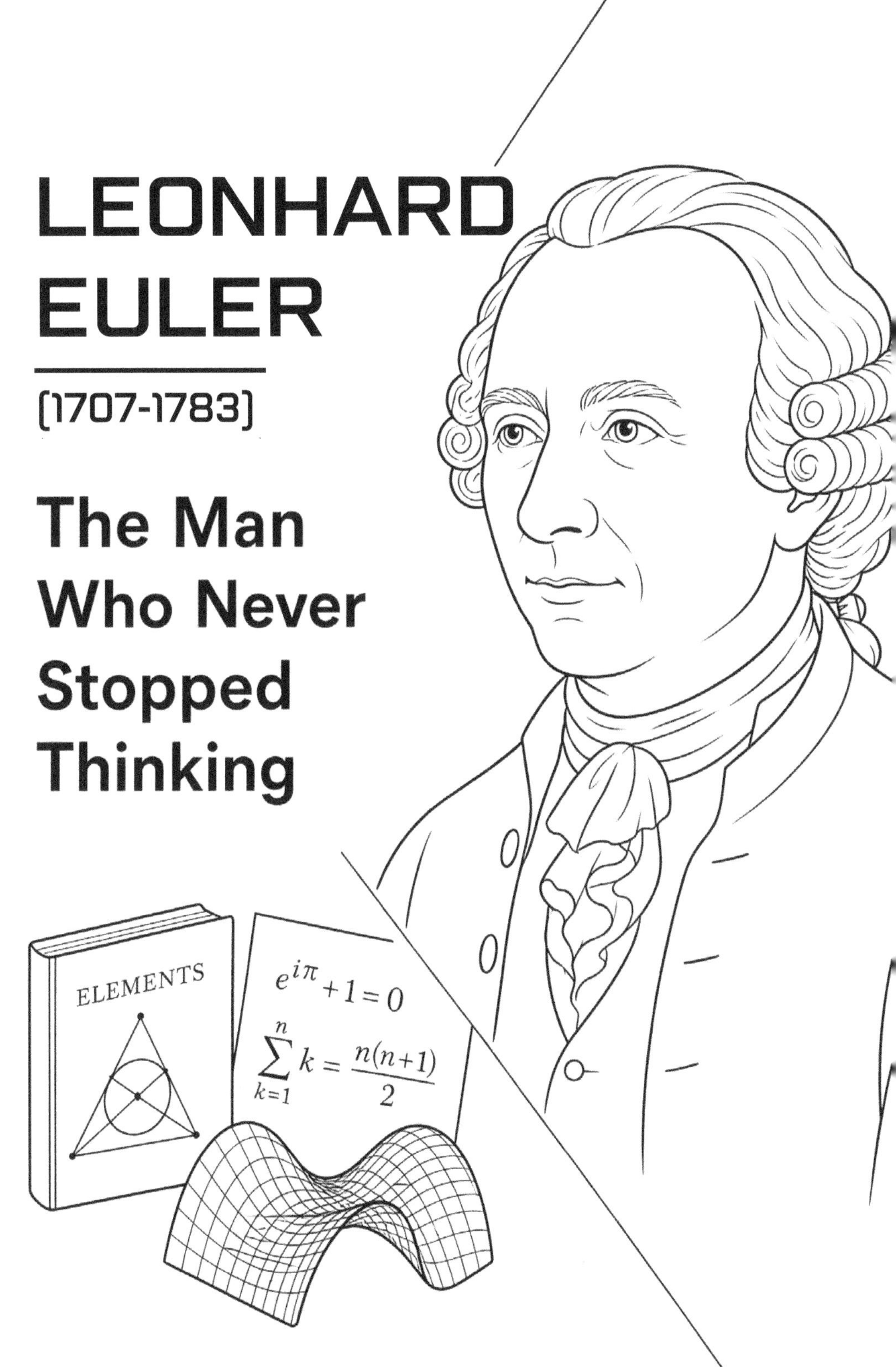

Your calculator has a button labelled "e." It represents a number called Euler's number: 2.71828... continuing forever without repeating. It appears in the mathematics of compound growth, population dynamics, radioactive decay, and the shape of a hanging chain.

Leonhard Euler put it there. Along with dozens of other mathematical discoveries so numerous that after he died, the Academy of Sciences in St. Petersburg kept publishing his new work for 47 more years because there was simply too much of it to release all at once.

He went blind at 59. His output increased.

Leonhard Euler was born in Basel, Switzerland, in 1707, the son of a Calvinist pastor who hoped his son would follow him into the ministry. Euler studied theology as his father wished, and mathematics on the side.

The mathematics quickly took over.

He had a kind of mind that saw patterns where other people saw randomness. Numbers were alive to him, connected and responsive in ways that other mathematicians found astonishing. He worked with a natural fluency that made his contemporaries look slow by comparison.

He moved to St. Petersburg, Russia, at the invitation of the newly founded Imperial Academy of Sciences, then to Berlin, then back to St. Petersburg. Wherever he went, he worked.

He worked with a focus and productivity that defied belief. He produced an average of 800 pages of mathematics per year across his entire career. He worked with children playing around his feet and never seemed distracted.

In 1738, he lost the sight in his right eye, possibly from overwork and eye strain. He kept going.

In 1766, he returned to St. Petersburg and within a few weeks suffered a cataract in his remaining eye. He went completely blind.

His mathematical output increased.

With no eyes, he held entire calculations in his memory with perfect clarity. He dictated his work to his sons and to students who came to assist him. He completed his full-length book on the motion of the moon from memory alone. He kept publishing original research until his death in 1783.

His faith was the quiet centre of everything. A devout Calvinist Christian, Euler led family Bible readings every evening and maintained what those close to him described as a deep, settled trust in God that expressed itself not in dramatic gestures but in daily faithfulness. He saw mathematical beauty as a glimpse of divine order.

He is estimated to have produced roughly a third of all the mathematical and scientific work produced in the entire 18th century.

Not in a flash of inspiration. In 800 pages per year. Every year. For decades.

"Though outwardly we are wasting away, yet inwardly we are being renewed day by day." - 2 Corinthians 4:16

Reflection and Prayer

Euler lost his sight and worked harder. What is the most significant obstacle you have faced in pursuing something you care about? What would it look like to refuse to let that obstacle define the limit of what you can do?

Lord, renew me daily. On the days when I feel like I'm losing ground, remind me that you are doing something in me that I can't always see. Help me to keep showing up, keep doing the work, and trust that the compound effect of faithful effort will eventually become something worth seeing. Amen.

Your Challenge

Try to memorise the first 15 digits of pi: 3.14159265358979. Give yourself 10 minutes. Write it out as many times as it takes. Euler memorised 100 digits for fun, after going blind. How far can you get? Come back tomorrow and try again. Notice what happens over three days of trying.

MAX
PLANCK
(1858-1947)
The Man Who
Accidentally
Changed
Physics
$E = h\nu$

Every LED screen. Every solar panel. Every laser. Every semiconductor chip inside every device you own. All of it works because of quantum mechanics, the physics of the very small.

Quantum mechanics began with Max Planck trying to solve a problem about the colour of glowing metal. He did not intend to start a scientific revolution. He was trying to fix a calculation that kept giving the wrong answer.

Max Planck was born in Kiel, Germany, in 1858, into a family with a deep tradition of academic and legal distinction. He grew up in a household where intellectual rigour and Lutheran Christian faith were both taken for granted and deeply intertwined.

He studied physics at the University of Munich and was famously warned by his professor that physics was essentially finished. All the major problems had been solved. There was nothing left to discover.

Planck became a physicist anyway.

He spent much of his early career studying thermodynamics, the physics of heat and energy. In the 1890s, he focused on a specific problem that had been quietly bothering physicists for years: blackbody radiation.

When you heat a piece of metal, it glows. It goes from dull red to bright orange to white as the temperature increases. The frequencies of light emitted at each temperature followed an observable, measurable pattern.

But when physicists tried to calculate the pattern mathematically using the existing laws of physics, the equations gave a wildly wrong answer. At high frequencies, the classical calculation predicted infinite energy output. This was so obviously impossible that physicists called it "the ultraviolet catastrophe."

Planck worked on it for years. He tried every possible approach within the existing framework. Nothing worked.

Finally, in 1900, he tried something desperate. What if energy was not continuous? What if it could only be emitted in discrete, specific packets, like coins rather than water?

He called these packets "quanta" (from the Latin word for "how much").

The mathematics worked perfectly.

Planck was deeply uncomfortable with what he had done. He believed in a clean, continuous, ordered universe. Quanta felt messy and arbitrary to him. He spent years trying to find a way to explain his result without quanta, hoping it was a temporary mathematical trick that would eventually be replaced by something more sensible.

It wasn't. Quanta were real. And what Planck had begun would eventually become quantum mechanics: the most accurate and extensively tested theory in the history of physics.

He received the Nobel Prize in 1918. His personal life had been devastated by two world wars: he lost his son Erwin to the Nazis in 1945 for alleged involvement in a plot against Hitler. He never recovered from that grief.

But he held his faith to the end, continuing to speak and write about the compatibility of science and Christianity until his death in 1947 at the age of 89.

"Whatever your hand finds to do, do it with all your might." - Ecclesiastes 9:10

Reflection and Prayer

Planck worked on one problem for years, tried every approach he could think of, and only found the answer when he was willing to try something that made him uncomfortable. Is there a problem in your life that keeps resisting the approaches you've tried? What would it look like to try something fundAmen.tally different?

Lord, give me Planck's willingness to sit with a problem long enough to actually solve it. Help me to resist the temptation to give up before the answer comes. And when the answer arrives in a form I didn't expect, give me the humility to accept it. Amen.

Your Challenge

Look at a candle flame, a light bulb, and the element of an electric cooker or toaster as they heat up. Each one glows a different colour at a different temperature. That colour difference is exactly the phenomenon Planck spent years trying to explain. Which is hottest: the red glow, the orange glow, or the white glow? (White is hottest. That is what Planck's equations eventually showed.)

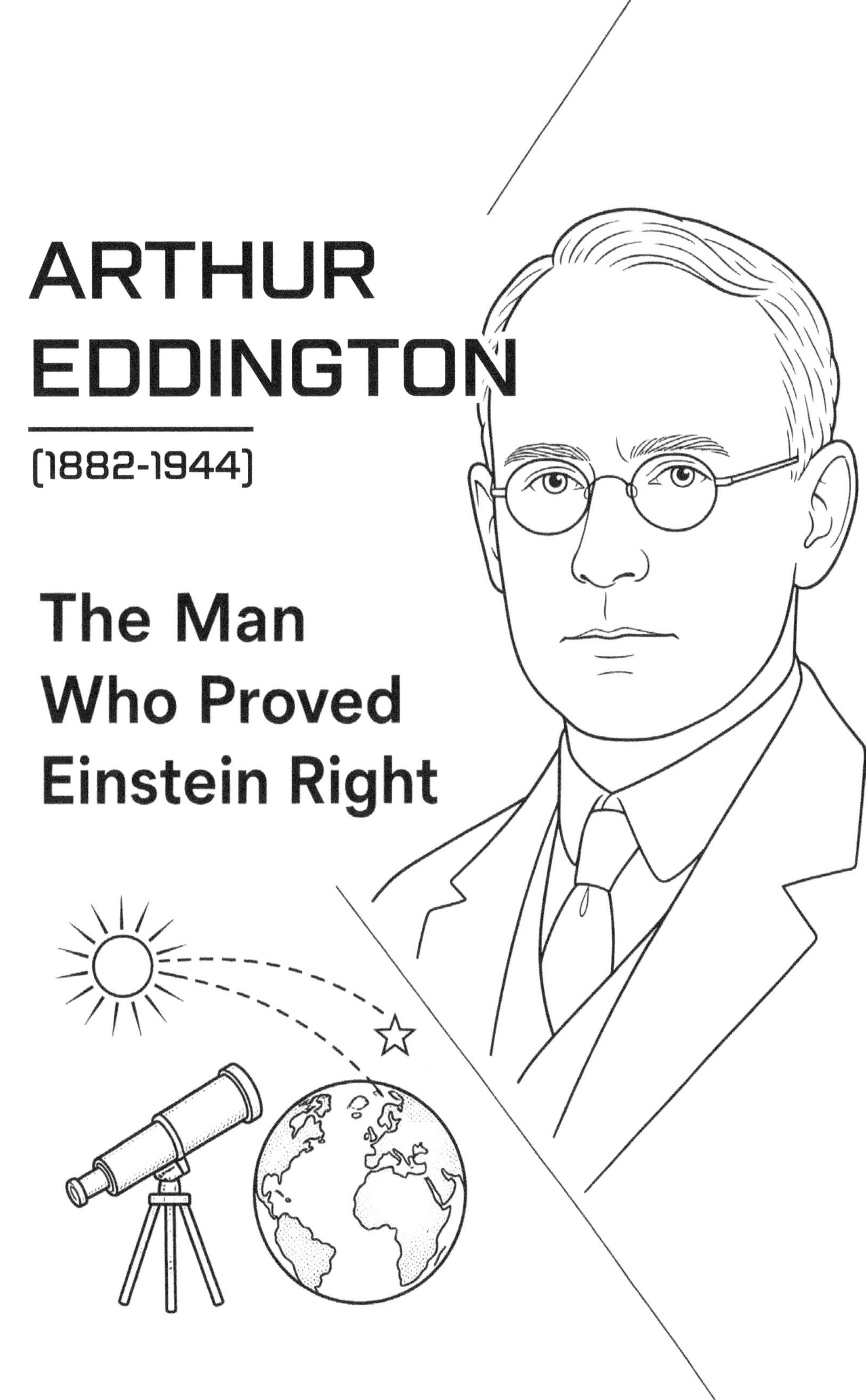
ARTHUR
EDDINGTON
(1882-1944)

The Man
Who Proved
Einstein Right

In 1915, Albert Einstein published his theory of general relativity: the idea that massive objects actually warp the fabric of space itself, bending it like a heavy ball placed on a stretched rubber sheet.

It was a breathtaking idea. It was also untested.

Arthur Eddington decided to test it. He waited. He planned. He sailed to a remote island in the Atlantic Ocean. He watched 302 seconds of darkness.

And he changed our understanding of the universe.

Arthur Eddington was born in Kendal, England, in 1882, into a Quaker family. His father died of typhoid fever when Eddington was two years old. His mother raised him and his sister in modest circumstances, making significant sacrifices to fund his education.

He was exceptional at mathematics from a very young age, won scholarships to Cambridge, and eventually became director of the Cambridge Observatory, one of the most prestigious positions in British astronomy.

He became one of the first scientists outside Germany to seriously engage with Einstein's new theory of general relativity. And he immediately grasped that there was a way to test it.

Einstein's theory predicted that light, which we normally think of as travelling in straight lines, would actually bend as it passed near a massive object like the sun. The sun's gravity, according to Einstein, would curve the space around it, and light would follow that curve.

If true, this meant that stars near the sun in the sky should appear very slightly displaced from their actual positions, because the light coming from them would have bent around the sun on its way to earth.

The problem: you can't normally see stars near the sun, because the sun is too bright.

Unless there is a total solar eclipse.

During a total solar eclipse, the moon blocks the sun completely, and for a few precious minutes, stars near the sun become visible. If Einstein was right, their apparent positions would be slightly different from where they were when observed at night, far from the sun.

Eddington had to wait for the right eclipse.

He planned two expeditions for the solar eclipse of May 1919: one to Sobral in Brazil, one to the island of Principe off the west coast of Africa, where he led the team personally. He had been working toward this moment for years.

On the day of the eclipse, clouds gathered over Principe. Eddington watched the sky with a sick dread, knowing that if the clouds didn't clear, years of preparation would be wasted.

The clouds thinned. For 302 seconds of totality, the sky was clear enough to photograph.

When the plates were developed and measured back in England, the stars were displaced exactly as Einstein had predicted.

Eddington stood before the Royal Astronomical Society in November 1919 and announced the result. The room understood immediately what it meant. The next morning, newspapers around the world declared that Newton's picture of gravity had been replaced.

Eddington was a lifelong Quaker whose faith was as important to him as his science. He believed that both science and religion were pointing at the same truth, approached from different directions.

He had waited years for 302 seconds. And every second had mattered.

"For the revelation awaits an appointed time; though it linger, wait for it; it will certainly come and will not delay." - Habakkuk 2:3

Reflection and Prayer

Eddington waited years for a few minutes of darkness. Is there something in your life that requires you to wait and prepare for an opportunity that hasn't arrived yet? What does faithful preparation look like when you can't control when the moment will come?

Lord, help me to prepare well for the opportunities I can't yet see. Give me Eddington's patience: the kind that uses waiting time productively, that keeps sharpening skills and deepening knowledge, trusting that the appointed moment will come. Amen.

Your Challenge

Look up when the next solar or lunar eclipse is visible from your location. Mark it on your calendar now, months or even years in advance. Then actually watch it when it comes. Eddington crossed an ocean for his. Yours requires only a calendar reminder. Preparation is the whole point.

ANTOINE LAVOISIER
(1743-1794)
The Man Who Rebuilt Chemistry From Scratch

Every time you breathe in, your body pulls oxygen from the air and uses it to release energy from the food you have eaten. Every time you breathe out, you release carbon dioxide as a waste product.

We know exactly what happens in that process, in precise chemical detail, because Antoine Lavoisier refused to accept an explanation that didn't quite make sense.

Antoine Lavoisier was born in Paris in 1743, the son of a wealthy lawyer. He was educated at some of the finest institutions in France and trained in law, as his family wished. But science occupied every spare moment he had.

He kept meticulous notebooks. Every experiment was recorded in full detail: what he used, how much, what he did, what he observed, and what the measurements showed. He was obsessive about precision in a way that was unusual even among scientists.

His question: what happens when something burns?

The accepted theory at the time was called phlogiston. The idea was that all flammable materials contained a substance called phlogiston, which was released during burning. When a candle burned out, it was because all the phlogiston had been used up. When metal rusted, it was because phlogiston was escaping slowly.

Lavoisier kept running into something that bothered him.

If phlogiston was being released during burning, substances should weigh less after burning. But when he burned certain metals, they actually got heavier.

That made no sense at all if phlogiston was leaving.

He designed a series of careful experiments to find out what was actually happening. He burned things in sealed containers and measured everything precisely: the weight of the substance before and after, the weight of the air before and after, the exact amounts of everything involved.

What he found was that burning was not a release of anything. It was a combination: a substance was reacting with a specific component of the air and forming a new compound. He isolated that component and named it oxygen.

He went further. He showed that water was not an element (as the ancient Greeks had believed) but a compound made of hydrogen and oxygen. He established the law of conservation of mass: in a chemical reaction, matter is neither created nor destroyed, only rearranged.

He gave chemistry its modern language: a systematic naming system for elements and compounds that is still used today.

His Catholic faith informed his approach to work. He believed in precision and honesty as moral virtues. Every measurement was a form of truth-telling. Every careful experiment was a form of respect for the created order.

In 1794, during the chaos of the French Revolution, Lavoisier was arrested on the basis of his previous work as a tax collector. He was tried, convicted, and guillotined in a matter of hours. The mathematician Joseph-Louis Lagrange said the next day: "It took them only a moment to cut off that head, and a hundred years may not give us another like it."

He was 50 years old.

His wife, Marie-Anne, had worked alongside him for years, translating scientific papers and illustrating his experiments. She preserved his notebooks and manuscripts and published his memoirs after his death.

His work outlasted the Revolution by centuries.

"Whoever can be trusted with very little can also be trusted with much."
- Luke 16:10

Reflection and Prayer

Lavoisier's greatness was built on the small, daily discipline of keeping precise records and measuring things carefully. Are there areas of your life where you are sloppy with small things? What would it look like to bring more care and precision to what you already do every day?

Lord, help me to be faithful in the small things. Give me Lavoisier's precision: the honesty to measure carefully, the discipline to record truthfully, and the patience to let rigorous daily work build into something significant. Amen.

Your Challenge

Light a candle (with adult supervision) and place a glass jar over it. Watch what happens. Lavoisier asked why the flame goes out when the air runs out, and his answer dismantled 2,000 years of accepted chemistry. The flame needs oxygen to burn. When the oxygen inside the jar is used up, the flame dies. That simple observation, measured precisely, changed everything.

CHAPTER 6

Collaboration: Standing On The Shoulders Of Giants

"If I have seen further, it is by standing on the shoulders of giants." - Isaac Newton

No one does this alone.

We have been taught to celebrate the lone genius. The solitary figure who locks themselves away and figures it all out by themselves. The dramatic moment of inspiration that nobody else was part of.

That is not how discovery actually works.

Every scientist in this chapter built on someone else's work. Every breakthrough in this book was made possible by predecessors who asked the first questions, by mentors who pointed the way, by colleagues who pushed back and made ideas sharper. Nobody started from zero. Nobody figured it all out by themselves.

Standing on someone's shoulders is not weakness. It is wisdom. It is how knowledge compounds across generations. One person's lifetime of work becomes the foundation for the next person's lifetime of work. Over centuries, humanity builds something that none of us could ever build alone.

The scientists in this chapter understood this better than almost anyone. They sought out giants. They wrote letters across continents. They read everything their predecessors had written. They gave credit. They passed the work forward.

Ask for help. Find your giants. Climb.

NICOLAUS COPERNICUS

(1473-1543)

The Man Who Moved the Earth

Every space mission ever launched. Every planet ever discovered. Every telescope ever pointed at the sky. All of it assumes that the sun sits at the centre of our solar system, not the earth.

Copernicus was the first person to put it there.

Nicolaus Copernicus was born in Royal Prussia (now Poland) in 1473, the son of a copper merchant. After his father died when Copernicus was ten, he was raised by his uncle Lucas, a Catholic bishop who believed deeply in the value of education. His uncle sent him to study at the University of Krakow and then to Italy, where Copernicus studied canon law, medicine, and mathematics.

Mathematics led him to astronomy. And astronomy led him to a question he could not let go of.

The accepted model of the universe, handed down from the ancient Greek astronomer Ptolemy and supported by the church for over a thousand years, placed the earth at the centre of everything. The sun, moon, planets, and stars all revolved around the earth in perfectly circular paths.

But the calculations never quite worked.

Astronomers had spent centuries adding complicated corrections to Ptolemy's model to make it fit the observations. The planets would sometimes slow down, stop, move backwards, and then move forward again. To explain this without moving the earth, astronomers invented a system of wheels within wheels called epicycles. The system became messier and more complicated with every generation.

Copernicus kept asking: what if we have been measuring from the wrong point? What if the sun is the centre?

He was not the first to ask this. The ancient Greek astronomer Aristarchus had proposed a sun-centred universe over 1,800 years earlier. Copernicus read him carefully. He stood on that ancient shoulder and went further.

He spent decades computing. He used only his naked eyes and geometric reasoning. He calculated, checked, recalculated, and rebuilt the model of the solar system from the ground up. He found that if you put the sun at the centre, the strange backward movements of the planets made perfect mathematical sense. No epicycles needed.

He was a Catholic canon and he knew his book would be controversial. He spent years circulating a summary privately before publishing the full work. He finally published "On the Revolutions of the Celestial Spheres" in 1543. The first printed copy reportedly arrived in his hands on the day he died.

He never saw the reaction. But the chain he started never stopped. Copernicus led to Galileo. Galileo led to Kepler. Kepler led to Newton. That chain is the spine of modern science.

His faith was a quiet constant throughout his life. He served the church as a canon for over forty years. He believed that studying the created order was an act of reverence toward the Creator.

He changed the universe by being willing to stand on a giant's shoulders and look further than the giant had.

"As iron sharpens iron, so one person sharpens another." - Proverbs 27:17

Reflection and Prayer

Copernicus did not start from nothing. He read Aristarchus, studied the astronomers before him, and built on 1,800 years of accumulated questions. Is there a mentor, teacher, or writer whose ideas have shaped how you think? What would it look like to take what they gave you and go one step further?

Lord, thank you for the people who have gone before me and left something worth building on. Help me to learn from them well, to honour their work, and to have the courage to take it somewhere they never reached. Give me humility to receive help and boldness to keep climbing. Amen.

Your Challenge

Look up how far the earth is from the sun: 1 AU, which equals 150 million kilometres. Now look up the distances of Mercury, Venus, Mars, and Jupiter from the sun in AU. Copernicus calculated all of these using only naked-eye observations and geometry. No calculator. No telescope. See if you can convert any of those AU distances into kilometres. He did this kind of arithmetic by hand. Every day. For decades.

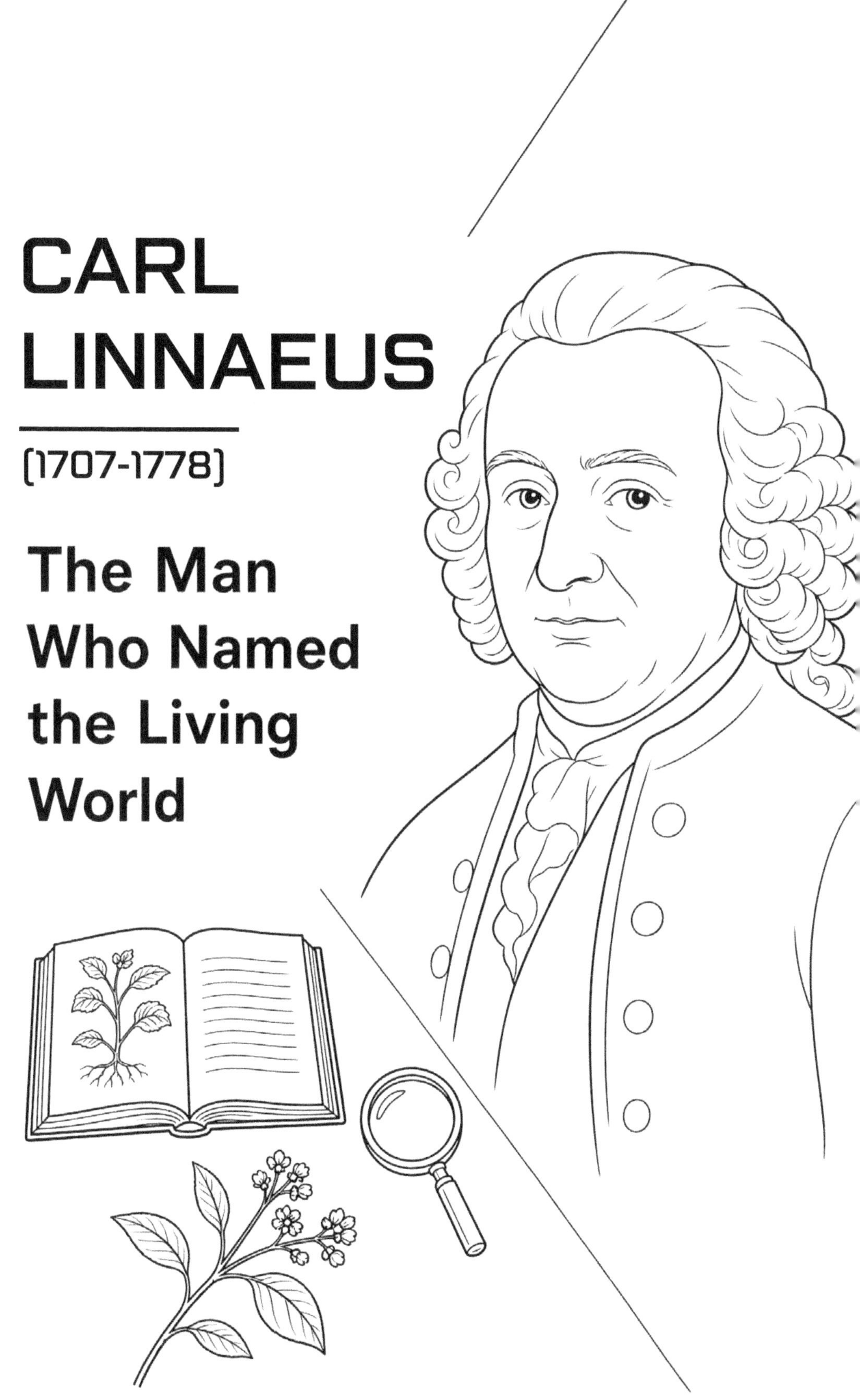

CARL
LINNAEUS
(1707-1778)
The Man
Who Named
the Living
World

Every animal encyclopaedia. Every biology textbook. Every time a scientist names a new species anywhere on earth. They are all using the naming system that Carl Linnaeus created, because he could not stand the chaos of how living things were being described.

Carl Linnaeus was born in Rashult, Sweden, in 1707, the son of a Lutheran pastor who kept a beautiful garden. He grew up surrounded by plants and became obsessed with them before he could read. His father taught him the Latin names of plants as a game when Linnaeus was very small, and the names lodged permanently in his memory.

He studied medicine and botany at the University of Uppsala and quickly noticed that the world of natural history was a mess.

The problem was this: a plant growing in England might have a completely different name in France, another in Germany, and another in Sweden. A bird described by one naturalist might be described again, under a completely different name, by another naturalist who had never read the first one. Scientists across Europe were studying the same creatures without knowing it. Knowledge was being duplicated, fragmented, and lost.

Linnaeus could not leave this alone.

His burning question was simple: is there an order beneath the chaos? Every living thing is different. But every living thing also belongs to something. Could one logical system describe all of them?

He was a deeply Lutheran Christian who believed that classifying nature was completing a task God had given to humanity at the very beginning. In Genesis, Adam is given the job of naming the animals. Linnaeus took this as a literal instruction. He believed that uncovering the structure of the natural world was a way of reading the mind of the Creator.

He did not start from nothing. He built on the work of English naturalist John Ray, who had already attempted a systematic plant classification. He read everything that had been published before him. He stood on those shoulders and pushed the system much further.

What he created was the binomial system. Every living thing would have two names. The first was the genus, a broader category it shared with its closest relatives. The second was the species, the specific name that belonged only to it. Homo sapiens. Panthera leo. Canis lupus. Two names. Every time. Every creature. Everywhere.

It sounds obvious now. It was not obvious then.

He named over 10,000 species of plants and animals himself. He sent his students across the globe to collect specimens, and seventeen of them died on those expeditions. He received thousands of specimens in letters and packages from naturalists on every continent. He classified them all.

His system gave science a shared language. Scientists in Sweden, England, France, and Japan could now all discuss the same species, using the same name, without confusion. The network of knowledge could finally be built properly.

The system Linnaeus created in the 1750s is still used, in refined form, by every scientist in the world today.

"He brought them to the man to see what he would name them; and whatever the man called each living creature, that was its name." - Genesis 2:19

Reflection and Prayer

Linnaeus believed he was doing sacred work by bringing order to the chaos of the natural world. He saw naming things carefully as a form of reverence. Is there an area of your own life where bringing order and precision could be an act of care, rather than just a practical task?

Lord, you made a world that is endlessly detailed and endlessly ordered. Help me to pay attention to that detail. Give me Linnaeus's patience for the small things, and help me to see that understanding the world you made is a way of knowing you better. Amen.

Your Challenge

Pick five objects in your room and create your own classification system for them. What categories do you use? Size? Material? Colour? Function? Write it out. Then try to add five more objects to your system without breaking the rules you just invented. Linnaeus did this with every living thing on earth, in a system that still holds 270 years later. How does your system hold up?

ASA
GRAY
(1810-1888)
The Man
Who Linked
Two Worlds

The fact that biology in America developed at all, that US universities teach plant science, ecology, and evolutionary biology, is largely because of one botanist who built a global network of relationships from a small office at Harvard.

Asa Gray was born in Sauquoit, New York, in 1810, the son of a farmer. He trained as a doctor and then abandoned medicine for botany, which paid almost nothing. He took a position as the first professor of natural history at the University of Michigan, then moved to Harvard, where he would spend the rest of his life building a botanical garden and one of the most important plant collections in North America.

He was not an obvious candidate for scientific greatness.

But he had a gift that was different from most scientists of his era. He was not just a brilliant observer. He was a connector.

Gray wrote hundreds of letters every year to botanists on every continent. He built relationships across Europe, Asia, and the Americas. He gave specimens, received specimens, asked questions, and shared answers. Long before the internet, he understood that knowledge moved fastest when people shared it freely.

His burning question came from something he kept noticing in his data. Certain plant species appeared in eastern North America and also in Japan, on the other side of the world, with no land bridge between them and no obvious way they could have crossed the Pacific Ocean. How had they got there?

He corresponded with Charles Darwin, and the two developed a working friendship long before Darwin published his famous book. Darwin trusted Gray enough to send him the first advance copy of "On the Origin of Species" before almost anyone else had read it. Gray reviewed it, defended it publicly, and argued that evolutionary theory was fully compatible with Christian faith.

Gray had not always been a Christian. He came to faith slowly as an adult, partly through the influence of his mentor John Torrey, a devout botanist whose household Gray had lived in when he was young. The faith he eventually held was thoughtful, not passive. He wrestled with the questions that Darwin's ideas raised and came out with his convictions deepened rather than shaken.

He saw no conflict between believing in God and believing that God might have used evolutionary processes. He wrote and lectured on this repeatedly, becoming one of the most prominent voices arguing that science and faith belonged together.

His plant question was eventually answered. Darwin's theory of common ancestry, combined with later geological evidence about ancient land connections and climate shifts, explained how species had dispersed so widely. Gray's careful data had helped point the way.

When he died in 1888, he had published over 350 scientific papers and books and had corresponded with virtually every serious botanist in the world. He had built bridges, not just between continents, but between people who would never have connected without him.

"Two are better than one, because they have a good return for their labor: if either of them falls down, one can help the other up." - Ecclesiastes 4:9-10

Reflection and Prayer

Gray's greatest contribution was not any single discovery. It was the network he built and maintained over decades. He made it his business to know people, share knowledge, and connect others. Who in your life could you reach out to, learn from, or connect with someone else who shares their interests?

Lord, help me to be a connector. Give me Asa Gray's generosity with what I know, his curiosity about what others know, and his patience to build relationships over the long term. Remind me that the most powerful thing I can do is sometimes not a discovery but a connection. Amen.

Your Challenge

Write a letter or an email to someone who knows more than you about something you are genuinely curious about. Ask them one real question. Be specific. Gray had hundreds of such correspondences running at the same time. One of them changed biology. You only need to send one.

WERNHER VON BRAUN
(1912-1977)
The Man
Who Reached
For The Moon

Every rocket NASA has ever launched. Every satellite now orbiting the earth. Every human being who has ever walked on the moon. All of it was made possible by engineering that Wernher von Braun built, standing on every rocket pioneer who had come before him.

Wernher von Braun was born in Wirsitz, Germany (now Poland), in 1912, into a wealthy and aristocratic family. He was a gifted musician and a restless student. He was also, from early childhood, completely consumed by a single idea: space.

He discovered the pioneers of rocketry as a teenager. He read the book "The Rocket Into Interplanetary Space" by Hermann Oberth, a Romanian physicist who had calculated mathematically that rockets could reach space. Von Braun read it as a teenager and immediately understood that this was what he wanted to do with his life.

His burning question was not really a question. It was a conviction: human beings could reach space. Oberth's mathematics proved it was possible. The only question was how.

He wrote to Oberth. He worked with Oberth. He joined an amateur rocketry club in Berlin, the Society for Space Travel, where engineers and enthusiasts were building and launching small rockets in an abandoned field. He devoured everything that American rocket pioneer Robert Goddard had published. When German authorities restricted access to Goddard's work, von Braun wrote to him directly.

He explicitly credited his giants. Goddard, Oberth, Konstantin Tsiolkovsky. Every step he took was built on what they had already proved.

During World War II, von Braun led rocket development in Germany under circumstances that were morally complicated and historically painful. After the war, he came to the United States, along with his team and his accumulated knowledge, and eventually became the director of NASA's Marshall Space Flight Center.

He became a born-again Christian as an adult. A Gideon Bible in a hotel room, given to him by a stranger, began a process that changed him. He spoke openly about his faith in his later years. He believed that the more he understood the scale of the universe, the more impossible it became to think that it had appeared by accident.

He said: the deeper you go into space, the more you encounter God.

He led the development of the Saturn V rocket, the most powerful machine ever built. On 20 July 1969, it carried three men to the moon. Von Braun watched from Mission Control. He had started as a teenager reading a book by a man he had never met.

He had climbed an enormous chain of shoulders to get there. And he never stopped crediting the giants below him.

"When I consider your heavens, the work of your fingers, the moon and the stars, which you have set in place, what is mankind that you are mindful of them?" - Psalm 8:3-4

Reflection and Prayer

Von Braun read Oberth. Oberth read Tsiolkovsky. Tsiolkovsky read Jules Verne. Every giant stood on another giant. Von Braun never forgot this. He gave credit freely and sought out knowledge everywhere he could find it. Is there someone whose knowledge, mentorship, or example has helped get you to where you are today? Have you thanked them?

Lord, thank you for putting people in my path who know things I don't. Help me to recognise my giants and be grateful for them. Give me the humility to ask for help, the curiosity to keep learning, and one day the generosity to become a giant for someone else. Amen.

Your Challenge

Look up the name Hermann Oberth. Then look up what Oberth's work owed to Konstantin Tsiolkovsky. Then look up what von Braun built on Robert Goddard. Draw a chain on a piece of paper, connecting each name to the next. Now trace a chain of people who have taught you things, from the earliest person you can remember all the way to now. Where does your chain start?

CHAPTER 7

Vision: Imagination Over Knowledge

"Imagination is more important than knowledge. Knowledge is limited. Imagination encircles the world."
Thomas Edison

Knowledge tells you what is.

Imagination tells you what could be.

Every scientist in this chapter saw something that did not exist yet. Not something they found lying around and studied. Something they imagined first, then chased until reality caught up with them. Some of them did not live to see it. Some of them were told it was impossible. Some of them were ignored for decades.

They imagined anyway.

God is the first visionary. He imagined a universe before there was anything at all, and then he made it. He spoke light into darkness. He breathed life into dust. When you imagine something new, something that does not yet exist, you are doing something that reflects the nature of the God who made you.

Think bigger. The universe is bigger than your current understanding of it. And it is waiting for people willing to imagine what has never been seen before.

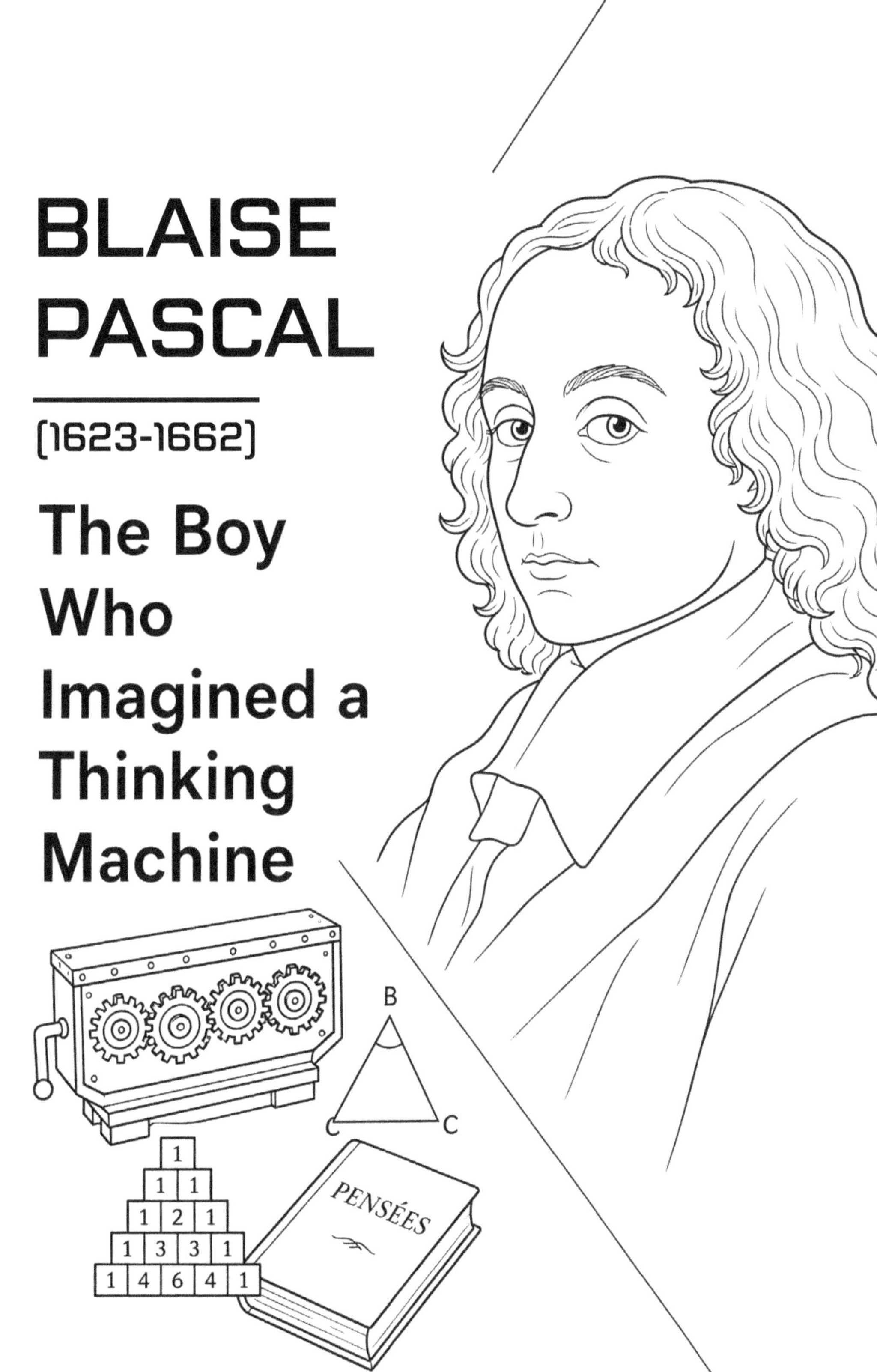

BLAISE
PASCAL
(1623-1662)
The Boy
Who
Imagined a
Thinking
Machine
B
C
C
PENSÉES
1
1 1
1 2 1
1 3 3 1
1 4 6 4 1

Every calculator. Every computer. Every phone. Every tablet. All of them trace back to a decision made by a French teenager who watched his father do exhausting arithmetic and asked: could a machine do this instead?

Blaise Pascal was born in Clermont-Ferrand, France, in 1623, the son of a brilliant mathematician and civil servant named Etienne Pascal. His mother died when he was three. He was raised by his father, who educated all his children himself and refused to let Blaise study mathematics.

Etienne believed mathematics was so addictive it would crowd out everything else. He hid the books.

It made no difference.

Blaise taught himself geometry in secret at the age of twelve, drawing shapes on the floor with charcoal because he had no paper and no books. By the time his father found out, the boy had already proved thirty-two propositions from Euclid on his own, including the theorem that the angles of a triangle add up to 180 degrees. Etienne reportedly wept.

The ban on mathematics was lifted immediately.

Pascal went on to write scientific papers as a teenager that impressed the most senior mathematicians in France. But what drove him most was a very specific problem he watched his father struggle with every single day.

Etienne Pascal had been appointed as a royal tax commissioner. His job required him to perform enormous amounts of arithmetic by hand: adding, subtracting, and checking columns of numbers for hours at a time. Mistakes were costly. Corrections were tedious. And there were no shortcuts.

Pascal could not stop thinking about it.

His burning question was not a theoretical one. It was deeply practical: what if a machine could do this instead? What if the mechanical regularity of arithmetic could be captured in gears and wheels?

He built fifty prototypes before he produced one that worked reliably. He was nineteen years old when he started and twenty-two when he finally solved it. The machine he created could add and subtract automatically. It was called the Pascaline.

It was never widely sold. It was expensive to manufacture and difficult for most users to operate. But the idea it contained, that a machine could do mathematical work, was the seed from which every computer that has ever existed eventually grew.

Pascal went on to develop probability theory with Pierre de Fermat, which became the mathematical foundation of statistics, insurance, and modern science. He made fundAmen.tal discoveries in fluid mechanics and atmospheric pressure that bear his name today.

He was also a deeply serious Christian thinker. He had a dramatic spiritual experience in 1654, a night of fire and certainty that he recorded and sewed into the lining of his coat. He carried it with him for the rest of his life. He wrote "Pensees" (Thoughts), one of the most penetrating and beautiful defences of Christian faith ever written.

He died at thirty-nine. He had imagined a computing machine a century before anyone built one, in a world where the fastest calculator was a human hand.

Now to him who is able to do immeasurably more than all we ask or imagine, according to his power that is at work within us." - Ephesians 3:20

Reflection and Prayer

Pascal saw a problem that everyone else accepted as just part of life, and he refused to accept it. He imagined a solution that did not yet exist and spent years building it. Is there a problem in your life or your world that you have been taught to accept as permanent? What would it look like to imagine a solution instead?

Lord, you are able to do more than I can ask or imagine. Help me to imagine bigger, because you are bigger than my best idea. Give me Pascal's stubbornness in the face of problems and his willingness to build fifty prototypes before finding the one that works. Amen.

Your Challenge

Add a column of twenty three-digit numbers by hand. Time yourself. Then use a calculator and time that. Pascal watched his father do the hand version thousands of times, for years, and couldn't accept that there was no better way. What repetitive task in your life do you think could be done better? Describe the machine or system you would build to replace it.

GEORGES LEMAÎTRE
(1894-1966)
The Priest Who Found the Beginning
$E = mc^2$

The Big Bang. The origin of the universe. The starting point of all space, time, matter, and energy.

It was proposed not by an atheist physicist in a laboratory. It was proposed by a Catholic priest.

Georges Lemaître was born in Charleroi, Belgium, in 1894, the son of a middle-class Catholic family. He studied engineering, then had his studies interrupted by World War I, during which he served as an artillery officer. After the war, he completed a doctorate in physics and was also ordained as a Catholic priest.

He was, simultaneously, a priest and a physicist. He saw no contradiction in this.

At Cambridge and then at MIT, he began working with Einstein's equations of general relativity, the set of equations that described how gravity worked across the entire universe. The equations were new, astonishing, and full of implications that physicists were still working out.

One implication bothered everyone, including Einstein himself.

The equations allowed for an expanding universe. If space itself was stretching outward in all directions, then the galaxies were moving apart from each other. Einstein found this conclusion so uncomfortable that he introduced a special correction factor into his equations, called the cosmological constant, specifically to prevent the universe from expanding. He called this constant his greatest blunder later in his life.

Lemaître read the equations without the correction.

His burning question was simple but enormous: if the universe is expanding now, what does that mean about the past? If you run the expansion backwards through time, what do you find at the beginning?

The answer was staggering. Everything, all of space, all of time, all of matter and energy, must have started from a single point. An extraordinarily small, extraordinarily hot, extraordinarily dense point. He called it the primeval atom.

Einstein heard the proposal and told Lemaître his physics was correct but his reasoning was "abominable." The idea that the universe had a beginning struck many scientists as too close to religious thinking to be taken seriously.

Lemaître kept going.

In 1929, the American astronomer Edwin Hubble published observations showing that distant galaxies were indeed moving away from each other, exactly as Lemaître's theory predicted. Einstein visited Lemaître at a conference in California, stood up in front of the audience, and publicly declared that what he had just heard was the most beautiful and satisfying scientific explanation he had ever encountered.

The beginning of the universe had been found. It was exactly the kind of beginning that a Catholic priest might have suspected was there.

Lemaître never thought his scientific work proved his religious beliefs. He kept the two separate because he thought that was the honest thing to do. He believed in God, and he believed the Big Bang was true, and he saw them as answers to different kinds of questions.

He died in 1966, having lived to see the discovery of the cosmic microwave background radiation, the faint afterglow of the Big Bang, which provided the final confirmation that his theory was right.

"In the beginning, God created the heavens and the earth." - Genesis 1:1

Reflection and Prayer

Lemaître proposed that the universe had a beginning when the established scientific consensus said it didn't. He held his position patiently, waited for the evidence, and was eventually proved right. Is there something you believe, based on careful reasoning, that the people around you are dismissing? What would faithful patience look like while you wait for the evidence to appear?

Lord, you were there at the beginning of everything. Help me to trust that you are also at the beginning of whatever I am trying to build. Give me Lemaître's patience to hold to what is true while the evidence catches up, and the humility to keep my certainties honest. Amen.

Your Challenge

Search for "cosmic microwave background" and look at the images. The faint glow visible in every direction of space is the afterglow of the Big Bang, the oldest light in the universe. Lemaître predicted it would exist before anyone knew how to detect it. It was confirmed in 1965, almost forty years after his original proposal. How long would you be willing to wait for your biggest idea to be taken seriously?

GOTTFRIED WILHELM LEIBNIZ
(1646-1716)
The Man Who Thought in 1s and 0s
STEPPED RECKONER
DYSERATIO DE ARTE COMBINATORIA
01000101
11010110
00101110
10011001

Every computer ever built runs on binary code. A system of 1s and 0s that can represent any number, any letter, any image, any sound, any instruction. Gottfried Leibniz invented it.

He also co-invented calculus. In the seventeenth century. In a Europe where most people still died of the plague.

Gottfried Wilhelm Leibniz was born in Leipzig, Germany, in 1646, the son of a moral philosophy professor. His father died when Leibniz was six, leaving behind a large private library. Leibniz was reading Latin and Greek by the time he was twelve, working his way through every book his father had owned.

He was one of the last people in history who could legitimately claim to understand everything.

He studied law, mathematics, philosophy, and physics. He invented a calculating machine. He served as a diplomat and court advisor. He wrote about theology, history, biology, and geology. He corresponded with virtually every significant thinker in Europe. He thought in multiple languages simultaneously.

His burning question was one of the most ambitious in the history of human thought: is there a universal logical language that could describe everything and resolve all human disagreement?

He believed that if you could find the right system of symbols and rules, you could settle arguments not by debate but by calculation. Any dispute, including theological disputes, could be resolved by working through the logic step by step.

His binary system was born partly from this vision. He read about ancient Chinese divination symbols called hexagrams, sequences of broken and unbroken lines. He saw in them a hint of what he was looking for. He developed binary arithmetic independently, a system where every number is expressed using only two symbols: 1 and 0. He was inspired partly by his Christian belief that God had created the world from something (1) and nothing (0).

He imagined that this system might one day be used in machines. Not the computing machines that would arrive 250 years later, but some simpler mechanical device. The full implications of what he had invented were far beyond anything anyone could yet build.

He also developed calculus simultaneously with Isaac Newton, working in Germany while Newton worked in England. The two men never met in person, and the bitter dispute over who had invented it first damaged Leibniz's reputation badly in his final years.

He died in 1716, relatively unrecognised and largely alone. His secretary was the only person to attend his funeral.

His binary system would eventually become the language of every computer ever built. His vision of a universal logical system would eventually inspire modern mathematical logic and computer science. His ideas outlasted the century by three hundred years.

Vision outlasts its visionary.

"I, wisdom, dwell together with prudence; I possess knowledge and discretion." - Proverbs 8:12

Reflection and Prayer

Leibniz imagined systems that would not be physically realised for 250 years. He died not knowing whether his biggest ideas would ever matter. Have you ever had an idea that seemed too big or too early to be useful? What does it look like to do the work faithfully, even when the results are far away?

Lord, you can see further than I can. Help me to trust that work done faithfully now can produce fruit long after I am gone. Give me Leibniz's courage to think bigger than my era, and the peace of knowing that whether I see the results or not, the work itself is worth doing. Amen.

Your Challenge

Convert your age into binary. (Search "decimal to binary converter" online.) Now convert the year you were born. Every number you will ever encounter can be expressed as 1s and 0s. Leibniz designed that system in 1679. Your phone runs on it today. What other things in your life are built on ideas from hundreds of years ago that you use without thinking?

WERNER HEISENBERG

(1901-1976)

The Man Who Found the Limit of Knowledge

$$\Delta x\, \Delta p \geq \frac{\hbar}{2}$$

$$\begin{bmatrix} p_x & p_y & p_z \\ q_x & q_y & q_z \\ L_x & L_y & L_z \end{bmatrix}$$

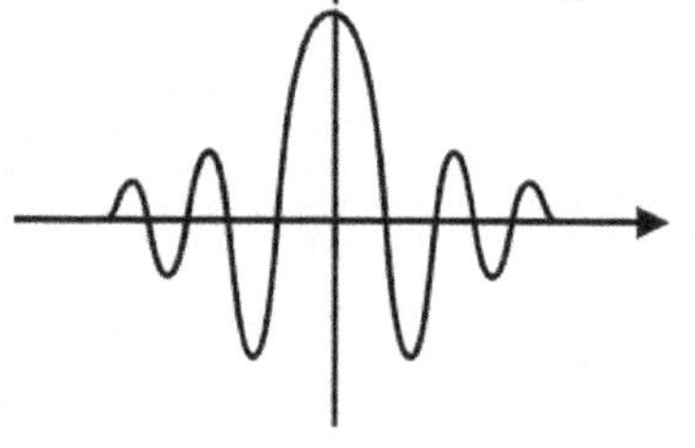

Every MRI machine that has ever scanned a human body. Every semiconductor inside every device you own. Every transistor in every computer, phone, or gaming console. All of it only works because of quantum mechanics.

And quantum mechanics is built, in part, on an idea that Werner Heisenberg found by staring at something that seemed impossible.

Werner Heisenberg was born in Wuerzburg, Germany, in 1901, the son of a Byzantine history professor. He was an excellent student and an enthusiastic outdoorsman who spent summers hiking in the mountains. He studied physics at Munich and went on to work with Niels Bohr in Copenhagen, one of the most exciting physics laboratories in the world in the 1920s.

The question that consumed him was one that the entire physics community was wrestling with: what were electrons actually doing?

Electrons, the tiny particles that orbit the nucleus of every atom, behaved in ways that classical physics could not explain. They absorbed energy and released it at precise frequencies. They jumped between energy levels in ways that seemed to have no gradual transition in between. Classical physics, the physics of Newton and Maxwell, described the behaviour of large objects beautifully. It completely failed for objects as small as electrons.

Heisenberg's burning question was this: when I try to describe exactly where an electron is AND exactly how fast it is moving at the same time, the maths refuses to give me both answers. Is that because my measurement is too crude? Or is something deeper going on?

He spent months turning this question over in his mind. The answer he arrived at was one of the most disturbing in the history of science.

It was not a measurement problem. The limit was built into reality itself.

The uncertainty principle states that the more precisely you know the position of a particle, the less precisely you can know its momentum (its speed and direction), and vice versa. This is not because our instruments are too clumsy. It is because the universe does not allow both pieces of information to exist simultaneously with perfect precision. Reality itself is fuzzy at the smallest scales.

Many scientists found this deeply troubling. Heisenberg found it philosophically beautiful.

He was a devout Lutheran who wrote and spoke extensively about science and faith throughout his life. He believed the universe was more mysterious and more wonderful than classical physics had suggested. He wrote: "The first gulp from the glass of natural sciences will turn you into an atheist, but at the bottom of the glass, God is waiting for you."

He received the Nobel Prize in Physics in 1932.

He imagined that reality was fundAmen.tally uncertain at its core, in a way that no amount of technology could ever eliminate. He was right. And that discovery made modern technology possible, because quantum mechanics, built on his uncertainty principle, describes the behaviour of the tiny particles that computers and MRI machines depend on.

"For my thoughts are not your thoughts, neither are your ways my ways," declares the Lord. "As the heavens are higher than the earth, so are my ways higher than your ways and my thoughts than your thoughts." - Isaiah 55:8-9

Reflection and Prayer

Heisenberg found that reality has a built-in limit on what can be known precisely. He didn't find this depressing. He found it beautiful. It reminded him that the universe is not fully transparent to human investigation. Is there an area of your life where you have been frustrated by not being able to know everything? What would it look like to hold that uncertainty with curiosity rather than anxiety?

Lord, your ways are higher than my ways. Help me to be comfortable not knowing everything. Give me Heisenberg's curiosity in the face of mystery, and remind me that the things I cannot fully understand are not threats but invitations to keep looking. Amen.

Your Challenge

Try this with a friend across a large room: attempt to clap your hands at exactly the same instant, with no signal, no countdown, and no looking at each other. The more precisely you try to synchronise, the harder it gets. That is not quite the uncertainty principle, but it gives you a small taste of why "exactly" is harder than it sounds, even before you get to the scale of electrons.

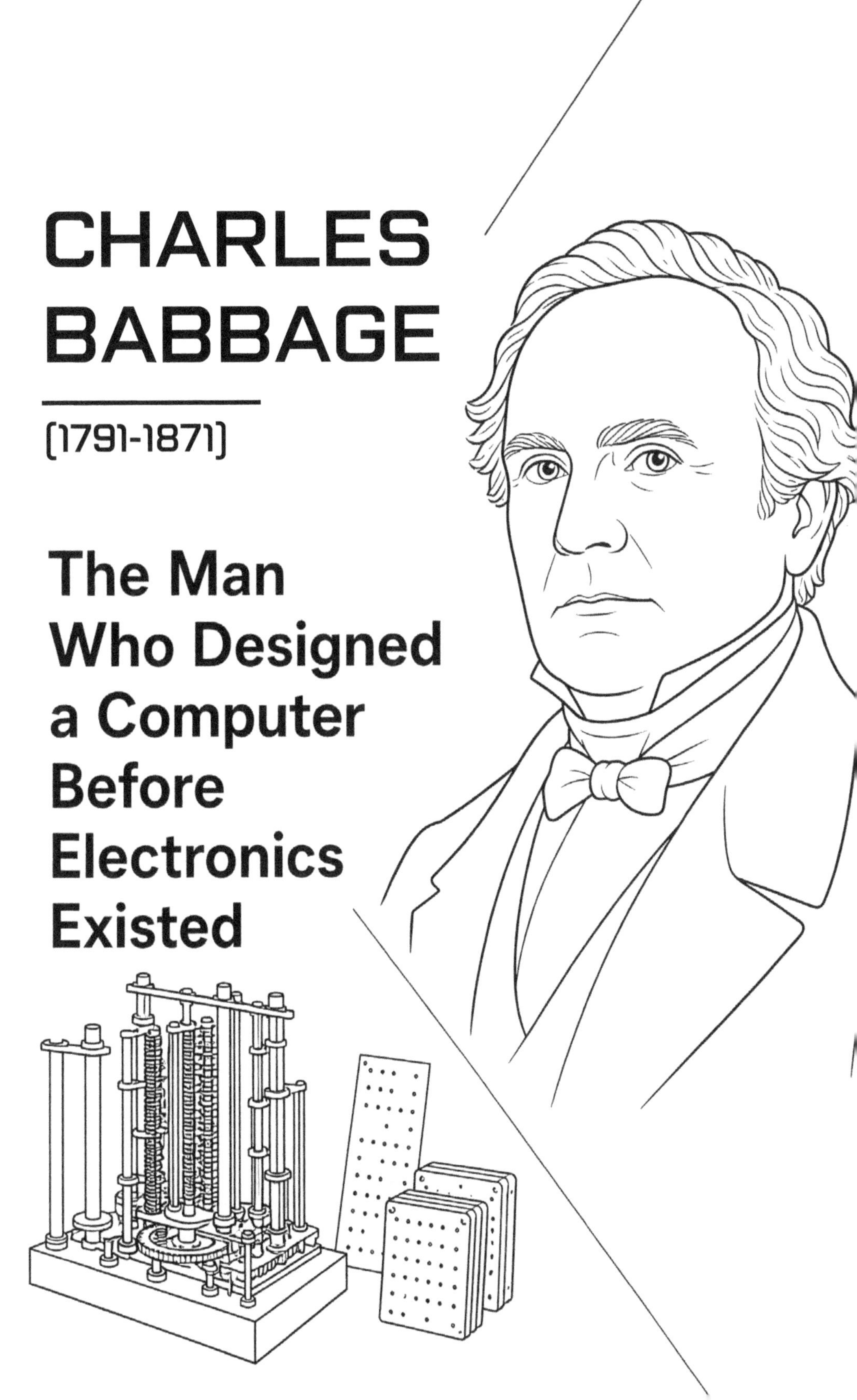

CHARLES
BABBAGE
(1791-1871)
The Man
Who Designed
a Computer
Before
Electronics
Existed

The laptop, phone, or tablet you are reading this on is the direct descendant of a machine that Charles Babbage designed in 1837.

One hundred years before transistors. One hundred and twelve years before the first electronic computer. In a world powered by steam.

Charles Babbage was born in London in 1791, the son of a banker. He was educated at Cambridge and became one of the leading mathematicians of his generation. He was brilliant, argumentative, endlessly curious, and perpetually frustrated.

He was frustrated by a specific problem that had consequences far beyond mathematics.

In the nineteenth century, the calculations used for navigation, astronomy, insurance, and engineering were stored in printed tables. Tables of logarithms. Tables of sines and cosines. Tables of tides and planetary positions. Ships relied on these tables to find their position at sea. Engineers relied on them to calculate stresses and loads. Errors in these tables could cause ships to run aground and bridges to fall.

Babbage checked the tables. They were full of errors.

Human computers (the people who calculated the tables by hand) made mistakes. The mistakes were copied into print. The print was used in the real world. People died.

His burning question was furious and practical: could a machine calculate these tables mechanically, without human error?

He designed a machine he called the Difference Engine, a mechanical calculator using gears and levers that could compute polynomial functions automatically, printing the results directly to avoid transcription errors. He convinced the British government to fund it. He spent years building part of it.

Then the project was cancelled. The engineering was too precise for the manufacturing methods of the time. The machine was never completed in his lifetime.

Babbage didn't stop. He designed something far more ambitious: the Analytical Engine.

The Analytical Engine was not just a calculator. It was a general-purpose computing machine. It had a "store" (what we now call memory), a "mill" (what we now call a processor), and the ability to follow conditional instructions: if this, then that. It could be programmed by punched cards. It could print its results. It included almost every architectural concept that modern computers use, designed in the 1830s using only mechanical parts.

It was never built.

Ada Lovelace, the daughter of the poet Lord Byron, studied Babbage's plans and understood them better than almost anyone. She wrote extensive notes on the machine, describing how it could be programmed to produce a sequence of numbers called Bernoulli numbers. Those notes are considered the first computer program ever written.

Babbage died in 1871, frustrated and largely unrecognised. He had spent over 17,000 pounds of his own money on machines that were never completed.

In 1991, the Science Museum in London built the Difference Engine No. 2 from Babbage's original designs. It worked perfectly. His vision had been exactly right. The world had simply needed a century to catch up.

He was a Protestant Christian who believed that God was the ultimate mathematician and that the order in the universe could be read through careful study.

He imagined a computer when the word didn't exist. He just had to wait for the world to get there.

"Now faith is confidence in what we hope for and assurance about what we do not see." - Hebrews 11:1

Reflection and Prayer

Babbage designed a computer that the world could not yet build. He was right. He just wasn't in the right century. Is there something you believe in or are working toward that feels ahead of its time? What would it look like to keep doing the work faithfully, trusting that the world will eventually catch up?

Lord, you can see the things I am building toward, even when I can't yet see how they will come together. Give me Babbage's stubborn belief in the work even when the results are not in sight. And help me to be someone who plants trees whose shade I may never sit under. Amen.

Your Challenge

T Find a logarithm table online. Pick any number and try to calculate its logarithm by hand using the rules of logarithms. Now imagine doing this for every number between 1 and 10,000, at ten decimal places of precision, by hand, making sure every single result is exactly right. That is the problem Babbage was trying to solve. Your calculator does it instantly. His machine would have done it in seconds. It took the world another 150 years to build it.

CONCLUSION

Now It's Your Turn!

You've met 30 people who changed the world.

Not because they were the smartest in the room.

Not because everything went their way.

Not because they had it all figured out before they started.

They changed the world because they couldn't stop asking

Faraday asked why a magnet moves when electricity flows, and gave us every electric motor on earth.

Nightingale asked why soldiers were dying after surgery, and gave us modern nursing.

Lemaître asked what Einstein's equations meant about the beginning of the universe, and gave us the Big Bang.

Pascal watched his father do arithmetic by hand and asked if a machine could do it instead, and gave us the computer.

None of them knew where the question would lead. They just followed it. Deeper and deeper, through failure and doubt and rejection and years of silence, until the answer came.

Here's what none of their textbook entries tell you:

Every single one of them had a faith that held when everything else fell apart.

Not a faith that made things easy. A faith that made things possible: it told them the universe was designed, that their curiosity was God-given, that the work mattered even when no one was watching.

That's the thing they all had in common. Did you see it?

Now here's the uncomfortable question this book has been building toward:

What's the question you can't stop thinking about?

Not the one you're supposed to have. Not the one that looks impressive. The one that wakes you up. The one you come back to when everything else is quiet. The one that feels too big, or too strange, or too specific to be useful.

That question is not an accident.

Mendel had peas. Faraday had magnets. Carver had a plant no one took seriously. You have whatever keeps nagging at you.

Follow it.

You don't need a lab. You don't need a degree. You don't need permission from someone who has already decided what you're capable of.

You need curiosity. You need the willingness to fail, and fail again, and call it data. You need the kind of faith that doesn't need everything to make sense before it takes the next step.

You have all of that. You were built with all of that.

"For we are God's handiwork, created in Christ Jesus to do good works, which God prepared in advance for us to do." - Ephesians 2:10

Prepared in advance.

Not improvised. Not accidental. Not dependent on whether you feel ready.

The work was prepared before you were. The question was placed in you on purpose.

Thirty scientists. Seven chapters. One idea running through all of it:

Genius doesn't start with an answer. It starts with a question you refuse to let go of.

You've seen what happens when people do that.

Now it's your turn.

Go find your question. Go deeper than is comfortable. Let your faith hold you when the answers don't come quickly.

And when they do come (and they will come), don't keep them to yourself.

The world is still waiting for what only you can find.

Review

If a story in this book stayed with you, if a scientist's struggle felt like yours, or if this book helped you see your curiosity as something God put there on purpose, would you leave a quick review?

Even one sentence goes a long way. As a small author, your feedback helps another young scientist or thinker find this book when they need it most.

 **Scan the QR code below** with your phone camera to go straight to the review page.

 **Or go to your Amazon orders** find this book, and tap "Write a product review."

**Thank you.
Genuinely.
It means everything.**

Author's Note

I wrote this book because I kept meeting teenagers who thought science and faith were enemies.

They had been told, or assumed, or somehow absorbed the idea that to believe in God meant you had to distrust science, and to trust science meant you had to give up on God. And so they lived in two separate boxes, slightly uncomfortable in both of them.

The thirty scientists in this book had never heard that false choice. They were Christians who studied the universe and found it more beautiful, not less, the more closely they looked. They were curious people who believed they were studying the work of a Creator, and that belief made them better scientists, not worse ones.

Faith gave them a reason to trust that the universe was real, rational, and worth investigating. It gave them resilience when the experiments failed. It gave them humility when they were wrong. It kept them going.

Your curiosity is not an accident. The capacity to ask questions, to wonder, to push back against easy answers: that is not a trait some people have and others don't. It is the image of God in you, working the way it was designed to work.

Use it. Ask the uncomfortable questions. Go deep into the hard problems. Build on the giants before you.

The world needs people willing to do what these thirty scientists did.

You are one of them.

Want to Go Deeper?

CHAPTER 1

Identity: Fearfully and Wonderfully Made

GEORGE WASHINGTON CARVER - George Washington Carver: From Slave to Scientist by Janet Benge and Geoff Benge (YWAM Publishing). Accessible biography written for young readers

MARY ANNING - Stone Girl Bone Girl: The Story of Mary Anning by Laurence Anholt. A picture book for younger readers, good for introducing the story - The Fossil Hunter by Shelley Emling. More detailed account for older readers

MICHAEL FARADAY - Michael Faraday: Spiritual Dynamo by Daryl Domning. Biography focusing on his faith and science - The Royal Institution website (rigb.org) has archives of Faraday's Christmas Lectures, some of which are still watchable online

FLORENCE NIGHTINGALE - Florence Nightingale: The Courageous Life of the Legendary Nurse by Catherine Reef

GREGOR MENDEL - Gregor Mendel: The Friar Who Grew Peas by Cheryl Bardoe. Excellent illustrated biography for younger readers.

CHAPTER 2

Resilience: 10,000 Ways That Won't Work

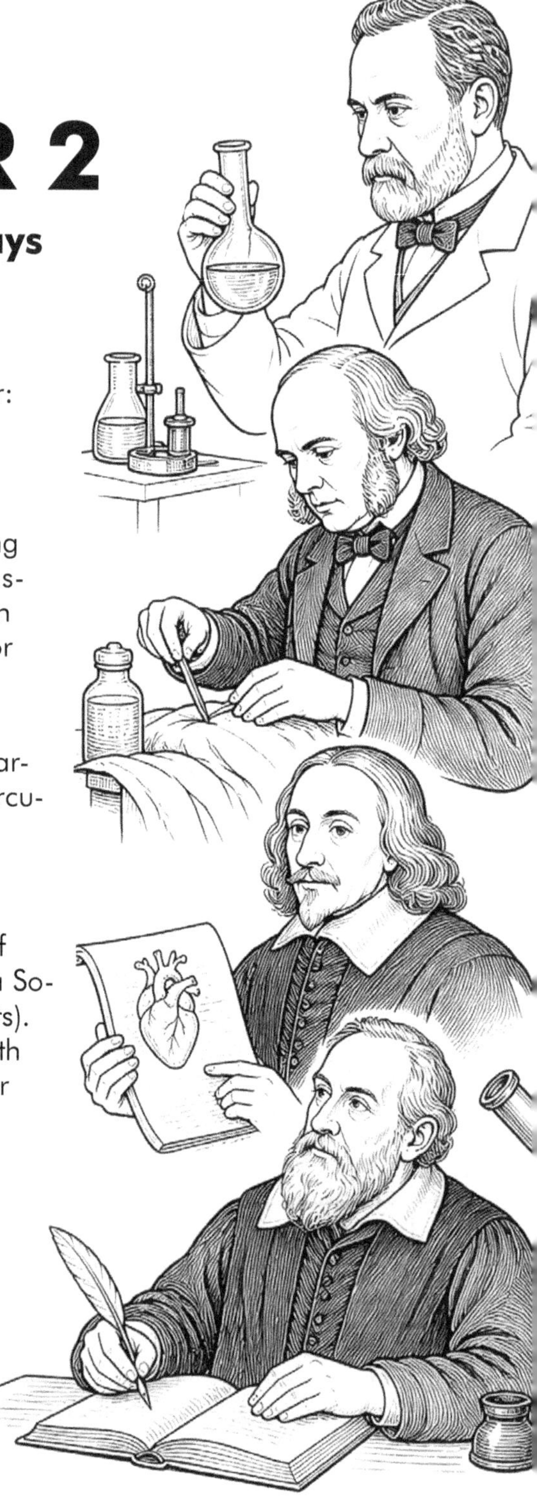

LOUIS PASTEUR - Louis Pasteur: Founder of Modern Medicine by John Hudson Tiner

JOSEPH LISTER - The Butchering Art: Joseph Lister's Quest to Transform the Grisly World of Victorian Medicine by Lindsey Fitzharris (for older readers)

WILLIAM HARVEY - William Harvey: Discoverer of How Blood Circulates by Rebecca Stefoff

GALILEO GALILEI - Galileo's Daughter: A Historical Memoir of Science, Faith, and Love by Dava Sobel (for older readers and parents). Beautiful account of Galileo's faith through his letters to his daughter

CHAPTER 3

Purpose: Think God's Thoughts After Him

JOHANNES KEPLER - Johannes Kepler: Discovering the Laws of Planetary Motion by Ann Gaines

ISAAC NEWTON - Isaac Newton: The Scientist Who Changed Everything by Philip Steele - Newton's theological writings are available online. His commentary on Daniel and Revelation reveal how deeply he took Scripture seriously.

ROBERT BOYLE - Robert Boyle: Trailblazer of Science by John Hudson Tiner

FRANCIS COLLINS - The Language of God: A Scientist Presents Evidence for Belief by Francis Collins. Collins's own account of his journey from atheism to faith, written accessibly for general readers

CHAPTER 4

Resilience: 10,000 Ways That Won't Work

JAMES CLERK MAXWELL
- The Man Who Changed
Everything: The Life of James
Clerk Maxwell by Basil Mahon

FRANCIS BACON - Francis
Bacon: Pioneer of Planned
Science by Doris Mary Stenton

CHARLES TOWNES - How the
Laser Happened: Adventures
of a Scientist by Charles
Townes. Townes's own account,
accessible and fascinating

LORD KELVIN - Energy, Force,
and Matter: The Conceptual
Development of Nineteenth-
Century Physics by P.M. Harman
(for older readers interested in
the physics context)

CHAPTER 5

Discipline: The Compound Effect

LEONHARD EULER - Euler: The Master of Us All by William Dunham. For older students who want to see the mathematics itself - The Euler Archive (eulerarchive.maa.org) contains many of his original papers

MAX PLANCK - The Quantum World: Quantum Physics for Everyone by Kenneth Ford. Good introduction to the physics Planck pioneered

Arthur Eddington - Arthur Eddington: Astronomer, Physicist, and Quaker by Richard Staley

ANTOINE LAVOISIER - The Disappearing Spoon and Other True Tales of Madness, Love, and the History of the World from the Periodic Table of the Elements by Sam Kean. Includes Lavoisier and many other chemists in a hugely readable format

CHAPTER 6

Collaboration: Standing On The Shoulders Of Giants

NICOLAUS COPERNICUS

- Copernicus: And the Revolution of the Celestial Spheres by Katharine Bosworth

CARL LINNAEUS -

Carl Linnaeus: Father of Classification by Margaret Jean Anderson

ASA GRAY - Asa Gray: American Botanist, Friend of Darwin by A. Hunter Dupree (for older readers)

WERNHER VON BRAUN

- Rocket Man: Robert H. Goddard and the Birth of the Space Age by David A. Clary. Tells the story of one of von Braun's giants - Carrying the Fire: An Astronaut's Journeys by Michael Collins. First-person account of the Apollo programme that von Braun made possible

CHAPTER 7
Vision: Imagination Over Knowledge

BLAISE PASCAL - Pensees by Blaise Pascal. His own writing, available in many translations. Even a few pages will show you how his mind worked. - The Phantom Tollbooth by Norton Juster. Not about Pascal, but captures the spirit of a mind that cannot stop asking questions

GEORGES LEMAÎTRE - The Day We Found the Universe by Marcia Bartusiak. The story of how the Big Bang was discovered and confirmed

GOTTFRIED LEIBNIZ - The Calculus Wars: Newton, Leibniz, and the Greatest Mathematical Clash of All Time by Jason Socrates Bardi

WERNER HEISENBERG - In Search of Schrodinger's Cat: Quantum Physics and Reality by John Gribbin. Excellent introduction to quantum mechanics for general readers

CHARLES BABBAGE - The Difference Engine by Doron Swade. The story of the Science Museum's project to build Babbage's machine from his original plans

Bibliography
Selected Sources by Scientist

George Washington Carver McMurry, Linda O. *George Washington Carver: Scientist and Symbol*. Oxford University Press, 1981.

Mary Anning Emling, Shelley. *The Fossil Hunter: Dinosaurs, Evolution, and the Woman Whose Discoveries Changed the World*. Palgrave Macmillan, 2009.

Michael Faraday Cantor, Geoffrey. *Michael Faraday: Sandemanian and Scientist*. Palgrave Macmillan, 1991.

Florence Nightingale Bostridge, Mark. *Florence Nightingale: The Woman and Her Legend*. Viking, 2008.

Gregor Mendel Iltis, Hugo. *Life of Mendel*. Translated by Eden and Cedar Paul. W.W. Norton, 1932.

Louis Pasteur Geison, Gerald L. *The Private Science of Louis Pasteur*. Princeton University Press, 1995.

Joseph Lister Fitzharris, Lindsey. *The Butchering Art: Joseph Lister's Quest to Transform the Grisly World of Victorian Medicine*. Scientific American / Farrar, Straus and Giroux, 2017.

William Harvey Frank, Robert G. *Harvey and the Oxford Physiologists*. University of California Press, 1980.

Galileo Galilei Sobel, Dava. *Galileo's Daughter*. Walker and Company, 1999.

Johannes Kepler Koestler, Arthur. *The Sleepwalkers: A History of Man's Changing Vision of the Universe*. Hutchinson, 1959.

Isaac Newton Gleick, James. *Isaac Newton*. Pantheon Books, 2003.

Robert Boyle Hunter, Michael. *Boyle: Between God and Science*. Yale University Press, 2009.

Francis Collins Collins, Francis S. *The Language of God: A Scientist Presents Evidence for Belief*. Free Press, 2006.

James Clerk Maxwell Mahon, Basil. *The Man Who Changed Everything: The Life of James Clerk Maxwell*. John Wiley and Sons, 2003.

Francis Bacon Jardine, Lisa. *Francis Bacon: Discovery and the Art of Discourse*. Cambridge University Press, 1974.

Charles Townes Townes, Charles H. *How the Laser Happened: Adventures of a Scientist*. Oxford University Press, 1999.

Lord Kelvin Smith, Crosbie, and M. Norton Wise. *Energy and Empire: A Biographical Study of Lord Kelvin*. Cambridge University Press, 1989.

Leonhard Euler Dunham, William. *Euler: The Master of Us All*. Mathematical Association of America, 1999.

Max Planck Heilbron, J.L. *The Dilemmas of an Upright Man: Max Planck as Spokesman for German Science*. University of California Press, 1986.

Arthur Eddington Stanley, Matthew. *Practical Mystic: Religion, Science, and A.S. Eddington*. University of Chicago Press, 2007.

Antoine Lavoisier Donovan, Arthur. *Antoine Lavoisier: Science, Administration and Revolution*. Blackwell, 1993.

Nicolaus Copernicus Gingerich, Owen. *The Book Nobody Read: Chasing the Revolutions of Nicolaus Copernicus*. Walker and Company, 2004.

Carl Linnaeus Blunt, Wilfrid. *The Compleat Naturalist: A Life of Linnaeus*. Viking Press, 1971.

Asa Gray Dupree, A. Hunter. *Asa Gray: American Botanist, Friend of Darwin*. Harvard University Press, 1959.

Wernher von Braun Neufeld, Michael J. *Von Braun: Dreamer of Space, Engineer of War*. Alfred A. Knopf, 2007.

Blaise Pascal Connor, James A. *Pascal's Wager: The Man Who Played Dice with God*. HarperOne, 2006.

Georges Lemaître Lambert, Dominique. *The Atom of the Universe: The Life and Work of Georges Lemaître*. Copernicus Center Press, 2015.

Gottfried Wilhelm Leibniz Antognazza, Maria Rosa. *Leibniz: An Intellectual Biography*. Cambridge University Press, 2009.

Werner Heisenberg Cassidy, David C. *Uncertainty: The Life and Science of Werner Heisenberg*. W.H. Freeman, 1992.

Charles Babbage Swade, Doron. *The Difference Engine: Charles Babbage and the Quest to Build the First Computer*. Viking, 2001.